T0296220

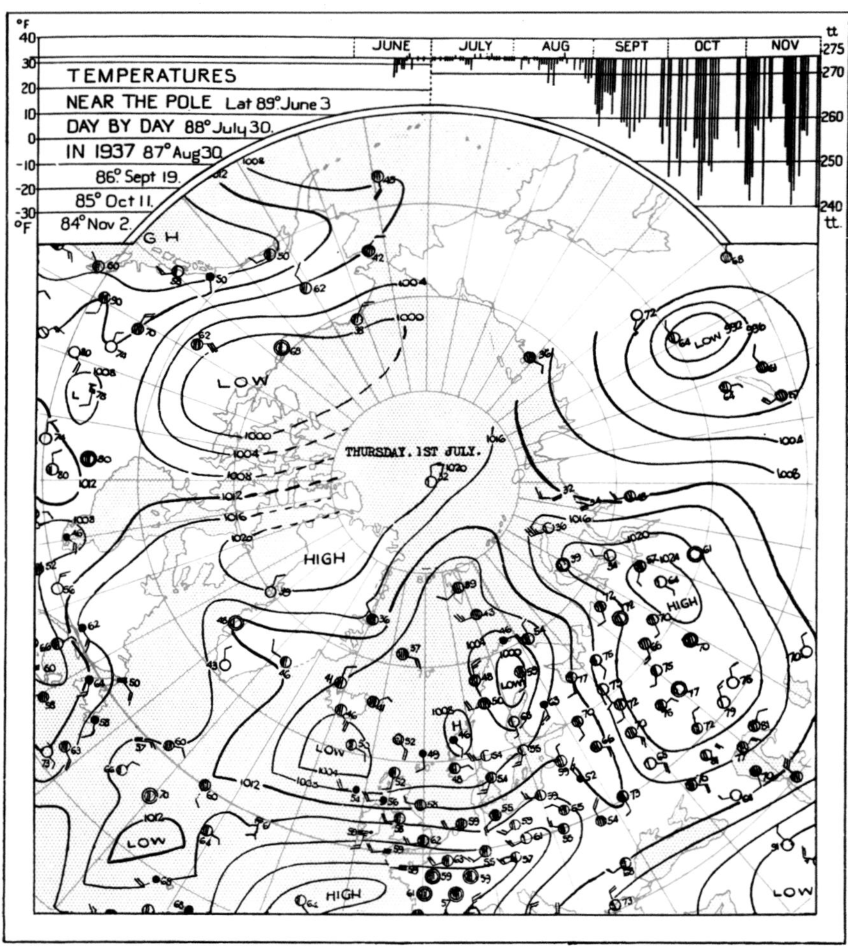

Summer weather in the North on 1 July 1937 at 1 a.m. (London Time) in the far west and 7 a.m. in the near west and east, with the Atlantic Low at 6 a.m. and records of temperature near the North Pole extracted from the Daily Weather Report of the Meteorological Office, with graphic conversion to tercentesimal "absolute".

By 19 February 1938 the drift of the ice-floe had carried the observers from the polar region to lat. 70° 47′ N., long. 19° 48′ W.

THE DRAMA OF WEATHER

THE DRAMA OF WEATHER

By

SIR NAPIER SHAW

Honorary Member and sometime President of the International Meteorological Committee

CAMBRIDGE

AT THE UNIVERSITY PRESS

MCMXL

CAMBRIDGE
UNIVERSITY PRESS

University Printing House, Cambridge CB2 8BS, United Kingdom

Cambridge University Press is part of the University of Cambridge.

It furthers the University's mission by disseminating knowledge in the pursuit of education, learning and research at the highest international levels of excellence.

www.cambridge.org
Information on this title: www.cambridge.org/9781107450875

First edition 1933
Reprinted 1934
Second edition 1939
Reprinted 1940
First published 1940
First paperback edition 2014

A catalogue record for this publication is available from the British Library

ISBN 978-1-107-45087-5 Paperback

TO
WATCHERS
OF THE WORLD'S WEATHER
NEAR AND FAR
THIS BOOK
OF UNCONVENTIONAL ESSAYS
IS
RESPECTFULLY AND GRATEFULLY
INSCRIBED

PREFACE

In 1914 a summer holiday was to have been devoted to the "filming" of the clouds as part of the study of weather. Camera and apparatus were ready—some preliminary efforts had promised success. But the War and its consequences intervened. The opportunity has not recurred and the apparatus has joined the ranks of the unemployed.

Yet the idea survives of a sequence of events in the sky as the progress of a drama quite as interesting for students of nature as some which achieve popularity on a smaller stage.

Without the aid of the kinematograph, this work is devoted to the development of the idea by methods which clothe the actors in the ordinary dress of weather-study. So may the accumulation of meteorological figures, like the dry bones of the valley, come to life again for their part in the pageant which has been transforming yesterday into to-morrow ever since the world began.

Now that astronomy has suggested that our knowledge of the greater universe of stars and planets points to a mathematical formula as its expression, there is room for the suggestion by the atmosphere that weather is the work of an accomplished dramatic artist; and, for its display, pictures are more potent than formulae.

Of the pictures upon which I am relying for the expression of this suggestion few can be called my own. I have always been and still am indebted to friends in many parts of the world for the kindness of their assistance.

In this present instance, for photographs, to Count Abe who has remarkable opportunities in Japan, Mr C. J. P. Cave of Stoner Hill, Hants, Mr G. A. Clarke of the Observatory, Aberdeen, Dr Vaughan Cornish of Camberley, a connoisseur of waves, Mr C. K. M. Douglas of the Meteorological Office, expert about the air and its ways, Commander L. G. Garbett, R.N., specialist in weather at sea (and with him Commander C. H. Lush, R.N., and Lieut.-Commander G. A. Williams, R.N.), Mr Robin Hill of Cambridge, Professor W. J. Humphreys of the Weather Bureau of the United States, Commandant J. Jaumotte, Director of the Belgian Meteorological Service, Señor R. Patxot i Jubert of Barcelona, a very good friend of meteorological science, Professor W. Kusnetzoff of Dorpat, Dr W. J. S. Lockyer of the Norman Lockyer Observatory, Professor A. McAdie, sometime Director of Blue Hill Observatory, Dr E. G. Mariolopoulos, Professor in the University of Salonica, Dr J. S. Owens, a well-known authority on atmospheric pollution, the Master of Sempill, at one time Aeronautical Adviser for Japan.

Nor can we forget the late Colonel F. A. Chaves, Director of the Meteorological Service of the Azores, Mr W. H. Dines of Benson, Oxon., Flight-Lieut. J. F. Lawson, R.A.F., stationed in Iraq, Mr A. Mallock of the old Advisory Committee for Aeronautics, Dr A. de Quervain of Zürich and Mr John Tennant, an enthusiastic stereographer.

To others, from whose published work I have borrowed, acknowledgment is made in the text and I may repeat once more the thanks recorded in the *Manual of Meteorology* to the Everest Committee of the Royal Geographical Society and to the Editor of *Nature*.

And last but not by any means least to H.M. Stationery Office for permission to reproduce figs. 35, 39, 46, 54 and 73 which have appeared in official publications and the weather-maps of figs. 81, 82, 86, 89 and 90; to the Meteorological Office for the loan of photographs for figs. 15*c* and 38 and for the original records from which fig. 44 was compiled, in addition to assistance with photographs previously reproduced in the *Manual of Meteorology*; to the Royal Society for permission to reproduce fig. 91, to the Royal Meteorological Society for the loan of the photograph for fig. 20, for the blocks of figs. 37, 65–7, and for permission to reproduce fig. 85 compiled from the *Quarterly Journal*; and to the Royal Astronomical Society for the loan of the photograph for fig. 25. The picture of the V-shaped depression is included in the frame of fig. 85 by permission of Messrs Kegan, Paul, Trench, Trubner and Co. Ltd., the publishers of Abercromby's *Weather*.

Recent activity in the study of the influence of weather on crops leads to the consideration of meteorological accumulations in relation to natural integration. Ordinary meteorology is essentially differential in its habit. In the weather-map we look for differences between this place and that, between day and night or between yesterday and to-day. When we come to to-morrow the past is over and done with except as a matter of history. And in our climatological tables we exhibit our means and extremes, and sometimes our totals, for the month or the year, and leave them to the reader's remembrance; but in deciding questions of growth and survival Nature never forgets the past, and therein we deal with integrals that the mathematician cannot emulate. That part of meteorology we call phenology. It has now an international organ of its own and for its use we must invoke gradual accumulations that do not appear on any weather-map.

The emergence of rhythm in certain aspects of weather is traced in the study of many records and the reader is introduced to some of the intrusions which the eternal rhythms have to suffer from day to day.

Due respect is inevitably paid to the weather-map as a help in answering the question, "Will it rain to-morrow?". It has thrilled humanity throughout the ages; but the reader will be forgiven if he finds that the text of the discourse points to the question, "Why did it rain yesterday?" as one which is worthy of attention.

Now that physics is so much concerned with the more recondite mysteries of radiation and radio-activity, with ions or electrons, protons or neutrons, the very word atmospheric has taken on a special electric meaning. The ordinary physical processes of the atmosphere related to the conservation of mass and energy and illustrated by "practical physics" which used to be the touchstone of the physical laws are in danger of losing the attention which their educative value deserves.

Just as the different aspects of what Aristotle called Meteorologics have now their own names and text books, as members of the commonwealth of the physical sciences, so, once more, some revision of the nomenclature is desirable in order that the physical processes of weather may continue to be recognised in the educational world as typical of the reactions that are continuous and conservative.

And meanwhile it is easy to recognise in the situation presented by the weather-map since its inception seventy years ago, first a barometer-solo as the leading motif of the Victorians and a duet for tropical and polar air in the Norwegian revision; and yet to feel that what nature desires to express is a symphony for temperature, pressure and wind, dramatised by the sun, led by the earth's rotation and conducted by gravity with water-power in reserve for the big side-drums.

The book relies for its information mainly on the four volumes of the *Manual of Meteorology* (Cambridge University Press 1926–32) and on *Forecasting Weather* (Constable & Co. Ltd. 1923). They were compiled with the assistance of Miss Elaine Austin, and Miss Austin's assistance has been continued in this work.

A subject that is built up round pictures is full of awkward corners for the printer; the simple process of turning over a new leaf sometimes puts one in the wrong street. It is only left to us to admire with gratitude the skill and patience with which Mr Lewis and his staff have managed to keep the author and the reader together on the high road.

NAPIER SHAW

London
10 *August* 1933

* * * * * *

With the passing of the year 1937 a revision of the Drama has been called for which necessitates some alterations and additions to the text and illustrations. It may be noted that scientific meteorology has been set in a new light by the publication in French and in German of a great work entitled "Hydrodynamique Physique avec applications à la météorologic dynamique" par V. Bjerknes, J. Bjerknes, H. Solberg, T. Bergeron. It is dated 1934 with a Preface by V. Bjerknes, Oslo, Octobre 1932 which sets out fully the features which are briefly referred to on p. 245 as the Norwegian duet for polar and tropical air with cyclonic accompaniment.

It has also been found desirable to deal rather differently with the representation of the data of weather in a form suitable for application to the effect of weather on the growth of crops and a separate chapter is devoted to that aspect of weather.

We have to add our acknowledgements for illustrations to H.M. Stationery Office for nine weather-maps in Chapter VI, and to the Topical Press Agency, Ltd. for the picture of the levanter of Gibraltar.

N.S.

19 *August* 1938

TABLE OF CONTENTS

PROLOGUE

PAGEANTRY IN THE SKY

We are as clouds that veil the midnight moon;
How restlessly they speed, and gleam, and quiver,
Streaking the darkness radiantly!—yet soon
Night closes round, and they are lost for ever.

* * * * * *

Man's yesterday may ne'er be like his morrow;
Naught may endure but mutability. P. B. SHELLEY

THE HUMAN INTEREST

From the beginning of civilisation the weather has been a subject not merely of curiosity but of vital interest and frequently of profound anxiety. None will deny that it is the weather that moulds the life of man. The human race may have realised its primary existence in the rank growth of torrid vegetation, needing no protection and no food but that which the forest or the jungle would naturally afford; but the first step towards agriculture and a permanent abode would need an appreciation of the conditions of weather and every step of progress would be just as dependent as the first on a knowledge of the ways of the weather.

Fig. 1. Aberdeen, 1919, July 21, 13h 12m. (Clarke.)

Primarily it must have been necessary to recognise the cycle of the seasons, so that reaping might follow sowing at the proper time; and the disastrous consequences of an insufficiency of rainfall or a destructive storm must have impressed the inexperienced husbandman in a manner which only pioneers in civilisation could fully appreciate.

Fig. 2. Aberdeen, 1919, July 21, 12h 42m. (Clarke.)

It is therefore not surprising that the initial stage of our civilisation should be disclosed in the regions where the sequence of weather is of the simplest, and yet the conditions of agricultural life are tolerable. The earliest records come from Babylon and Egypt, countries where the alternation of warmth and cold is almost as regular as the motion of the stars; where rain is not a primary necessity because the river can be trusted year in and year out to provide a sufficient supply of the water upon which all life depends.

In the description of the Chaldean paradise of the Garden of Eden we find no guarantee of regular rainfall. Four rivers supply it with water. Herodotus tells us that the Egyptians of his time were very scornful of the Greeks who were dependent upon the goodwill of their gods for the supply of water, by rain from heaven, instead of having a constant supply laid on by river. For the Egyptians, the constant dependence of the inhabitants of Greece, and still more of those of Palestine, upon rain from heaven is more suggestive of purgatory than paradise. Indeed, it is difficult to find any phase of human existence so suggestive of demands on man's ingenuity for the protection of himself and his belongings, and the choice of suitable crops, as that of the dwellers in what are called temperate latitudes with variable winds and fluctuating weather. In spite of lean years there was corn in Egypt when Syria was famine-stricken. The ancient Egyptians studied their Nile as carefully as we study our rainfall.

We may carry a stage further the metaphorical association of ideas by remarking that the Greeks, a hardy race coming southward from the northern purgatory, regarded the country beyond Egypt as an inferno, "a white man's grave", where life was impossible on account of the heat.

Fig. 3. Aberdeen, 1921, May 13, 7h 0m. (Clarke.)

THE PAGEANTRY OF WEATHER

What then is this weather upon which all life on the planet depends for its existence? It is an ever-changing scene in which the clouds may be said to take the "star" parts, the parts indeed that the heavenly bodies take in the companion study of astronomy. Let me express my meaning by my personal experience. In front of my writing-table is a panel of southern sky bounded by a narrow window-frame; on the left-hand side is a spire with a weather-cock on the top projecting above a row of trees from a circle of stone angels. The trees reach up to 10° from the horizon, the spire to 20°, the panel to 60°, in effect it is not unlike the opening panel of the dome of an observatory. The bit of the sky that is bounded by the window-frame is a fair representation of a stage. There are occasions on which the sky is serene and the stage is empty, but generally there are clouds to be seen; they move, they change—what endless varieties of them there are—all on business, with a purpose in their direction and behaviour. The stage may show three or four different types at the same time layered one above the other, each layer with its

Fig. 4. Aberdeen, 1921, February 25, 8h 10m. (Clarke.)

own business and its own activity. Look at the sky yourself and see what actors are on the stage; turn away your eyes for a minute and then turn them back to see if you can recognise the actors that were there before. You will nearly always be impressed by the change. One cloud is increasing, another vanishing. At one moment the intervals between clouds are blue, after a minute nearly white. There are all shades of gradation of brightness or shadow, each one of which has a meaning.

The whole stage is full of action, the action which carries on the life of the world, for clouds are the precursors or the survivals of the rainfall upon which the world thrives. Even the waters of Egypt and Babylon are in fact dependent on the rainfall on distant mountains well situated for supplying the water which is otherwise denied to the plains. Broken cisterns in the ruins of Palmyra and other relics of civilisation of the Middle East are eloquent of the dependence of humanity upon water.

Meteorologists illustrate their books of instructions to observers with photographs of clouds, a dozen or more separate varieties, based upon the classification devised by Luke Howard a century ago—*cirrus*, the thread-cloud, *cumulus*, the heap-cloud, *stratus*, the sheet- or layer-cloud, and *nimbus*, the rain-cloud. Each is a kind of soliloquy which tells its own story but does not fully represent what nature generally provides. The occasions have to be chosen when the separate types can be isolated in the camera. On ordinary occasions the clouds of simple type are mingled with others in the aggregate of the sky.

So a true picture of clouds becomes very suggestive of the play of wind and weather to which we would direct the reader's attention, using for that purpose four streaks of sky photographed by Mr G. A. Clarke of the Observatory of Aberdeen (figs. 1–4). There is some difficulty about photography because precautions have to be taken to prevent blue sky appearing blank white in the picture; and in avoiding that, the blue sky may be so darkened as to appear black. But a photographic camera, or its predecessor the camera obscura which throws a picture of the sky on a table for the close inspection of the student, is the best introduction to the study of weather, although its field of view, being only as long as it is broad, may contain less information than the panel of vision of a long narrow window.

THE CAMERA, THE STEREOSCOPE AND THE FILM

Fig. 5. Southerly buster. (Commander C. H. Lush, R.N.)

As an example of the curious things which the camera can see we give a picture (fig. 5) of a "southerly buster" bringing cold squalls to Australia, taken at sea not far from Sydney Harbour at 6.15 p.m. on 25 January 1932 and reported as 20 miles long, the top 2600 ft and the bottom 800 ft.

Still better than the view of a single camera is the stereoscopic view provided by a pair of cameras a suitable distance apart. The combination gives a surprising amount of information that the single camera does not disclose.

We give two examples, one (fig. 6 *a*) by the late Mr John Tennant which shows the structure of the underside of a fracto-cumulus cloud, and the

Fig. 6 *a*. Irregular clouds. (Tennant.)

Fig. 6 *b*. Cumulo-nimbus, Barcelona. (Patxot.)

other (fig. 6 *b*) from Barcelona which gives us an insight into the influences of the lower reaches of the sky.

And the art of photography of the sky has reached its climax in the stereo-photographs of the whole sky which are provided by a special form of camera designed at Cambridge some years ago by Robin Hill. A specimen is reproduced here (fig. 7). Truly the stereoscope has lost its social status in the drawing-room, but we do not hesitate to ask the reader's aid in retaining it for the meteorological observatory.

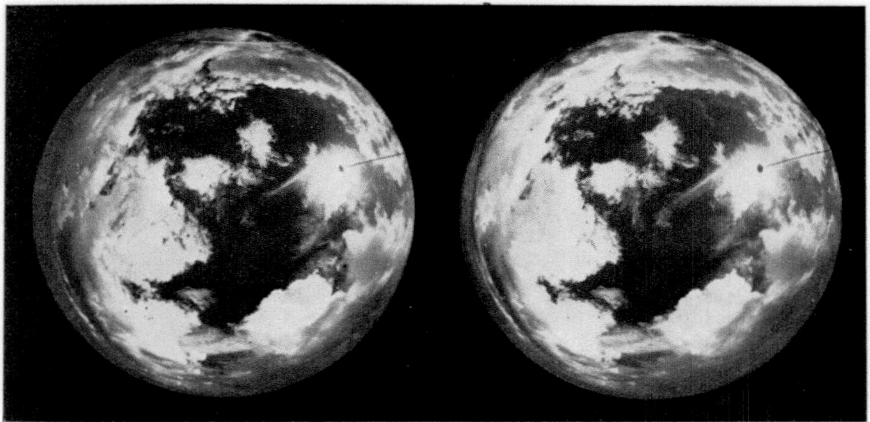

Fig. 7. Cambridge, 1924, July 29. (Robin Hill.)

For the stereoscopic photographs which are reproduced in this book the reader will find it convenient to use the pair of lenses of a stereoscope detached from the case which usually carries them. He will have a little trouble at first in superposing, one upon the other, the proper pair of pictures to give the stereoscopic effect; the distance of the photograph, and the angle between the line of the two photographs and the line of the two lenses have to be adjusted, but the result is worth the trouble.

The photographs that we have reproduced are sufficiently suggestive of the pageantry of the sky and we give them with the hope that some day the play of wind and weather may find itself represented on films which will disclose the life-history of the players. It is so transitory that it eludes precise description by the unaided spectator.

Fig. 8 *a–d*. Eruptive clouds.

a. Mount Asama, Japan.

b. Hampshire Downs, 1923, July. (Cave.)

c. Java, 1921, April 22. (C. Braak.)

d. France, 1918, Sept. 27. (Douglas.)

SOME SCENES OF THE DRAMA

The camera has enabled us to introduce to the reader the pageantry of the atmosphere as it might be viewed from a study window; and here we follow that introduction by a collection of pictures of various special aspects of the play that will help us in the further study of weather.

Our first steps will be to suggest the possibility of classifying the various features of the pageantry by exhibiting some photographs of the different forms of cloud, and then to add some views of various things that have transpired and may at some time or other come within the cognisance of anyone who carries a camera.

Let us take the items in order and provide the reader with a catalogue of the exhibition.

Eruptive clouds

The four pictures on p. 8 represent or at least suggest eruptions of different kinds but not dissimilar in form.

Fig. 8*a* shows the cloud of débris thrown up by the eruption of a volcano in Japan, and it is worth while to notice the analogy between the shape of the cloud and that of a tree of thick foliage in the first stages of full leaf. The analogy is preserved perhaps not quite so obviously in the other three pictures of which fig. 8*b* represents the cloud of an approaching thunderstorm in the south of England. Such clouds carry the name of cumulo-nimbus meaning a heap or cumulus cloud that is also a rain-cloud (nimbus) and implies a shower. Of all the typical clouds they are the thickest; there may be thousands of feet or even of metres between the base which is generally about 2000 ft up and the top which may be at 15,000 ft or higher. They often end in a curious fringe of thread-like cloud, stretching out like pointing fingers which carry the name of hybrid cirrus because they are fibrous in appearance but do not float separate in the sky. There are some examples on p. 10.

Fig. 8*c*. A cloud with exactly similar characteristics in Java enables us to call attention to the similarity of cloud-forms all over the world.

Fig. 8*d* shows eruptive clouds of the cumulo-nimbus type as seen from an aeroplane; they reach up to a height of 2000 metres (6500 ft) from the upper surface of a layer of lower cloud.

Fig. 9 *a–f*. Thunderclouds and rain.

a. Hampshire Downs, 1925, May 17. (Cave.)

b. Filey, 1925, March 22. (Cave.)

c. Hampshire Downs, 1924, May 19. (Cave.)

d. Aberdeen, 1920, April 7. (Clarke.)

e. Hampshire Downs. (Cave.)

f. Aberdeen, 1919, October 2. (Clarke.)

Thunderclouds and rain

On p. 10 are six pictures. The first, fig. 9*a*, shows alto-cumulus castellatus, cloud-heaps (at a level of perhaps 2 km, 6500 ft) with eruptive turrets, castles in the air, not unlike those of fig. 8*d* extending upward for some kilometres. They are a very sure sign of thunderstorms to come.

Fig. 9*b* shows a long layer of alto-cumulus with nimbus beneath; a massive heap-cloud above it with the pointing fingers of hybrid cirrus extending to right and left. Heavy rain is in progress underneath. It was taken at sunrise, and quite appropriately is over the sea, for sunrise is not at all a common time, indeed a very uncommon time, for thunderstorms inland.

Fig. 9*c* is also a picture of a distant mass of thundercloud 100 kilometres, 60 miles away, with a layer of fibrous cloud above it. The fact that it is visible from so far implies a great height for its top.

Figs. 9*d* and *e* show the last stages of a thunderstorm. One gives the ragged edge underneath the thundercloud while the rain is still falling, and the other the pocky surface (mammato-cumulus) that is seen when the storm is over.

Fig. 9*f* is entitled to its position as the fitting end of a shower. It represents what the observer sees when he has the sun at his back and drops of water in front of him with a cloud for dark background against which the rainbow can be seen. The primary bow shows colours, red uppermost, outside are the fainter colours of the secondary bow and inside supernumerary bows. A raindrop analyses the white light of a ray from the sun into its constituent colours; the most brilliantly coloured part of the primary bow forms a ring that lies at an angle of 42° with the line from the sun to the observer, the secondary bow at 51°. So if the sun is on the horizon, just risen or about to set behind the observer, the bows will be semicircular arches of 42° and 51°. If the sun is higher a smaller part of the ring will be visible, the feet of the arches will be lost; until when the sun is at 42° only the top of the primary can be seen on the horizon, and at 51° only the top of the secondary. So in temperate regions it is vain to look for rainbows in summer when the sun is high in the middle of the day, even if sun and rain are on the stage together.

Fig. 10 *a–c*. Disorder and quietude.

a. Aberdeen. (Clarke.)

b. North Atlantic, 1922, February. (Cave.)

c. Hampshire Downs, 1923, November 25. (Cave.)

Disorder and quietude

The next illustrations of the pageant show in fig. 10*a* at the top of the page a view of groups of fibrous cloud, called cirrus, suggestive of a confused whirl of tufts of white hair. It comes from Aberdeen and belongs to the central region of a summer anticyclone under which, indeed, at the earth's surface, as will be explained later, we should expect the quietude of a warm summer day.

Meteorologists say that these threadlike clouds are not composed of water-drops but of minute ice-crystals, and aviators confirm that conclusion.

Whether the form of the cloud really indicates turmoil at its own level or is just the expression of local differences in the load of water which the air is carrying may be a matter for future inquiry. Nature is not at all unwilling to suggest things that turn out after all not to be true. One of the primary lessons for a student of weather is not to believe all that you think you see.

The other two pictures on the page certainly suggest quietude. Fig. 10*b* with the dark clouds, which, if we can recognise it, is a broken up layer, we will call irregular stratus. Mr Cave brought this from the Atlantic, north-east of Madeira. The darkness and smoothness of the cloud belongs to the quiet of an evening sky. The warmth of the day is over and the masses of cloud drifting from the west gradually lose themselves.

At the bottom of the page, fig. 10*c* shows on the other hand the quietness of early morning in a Hampshire valley which, left undisturbed the night through, has filled itself with fog the level top of which looks not unlike the surface of a great lake.

Valley-fog is a very interesting phenomenon which will tax the ingenuity of anyone interested in the physical processes of weather. It is easy to suggest that under a clear sky at night the ground is cooled because its heat is lost by radiation into space; that in itself would entitle it to a liberal allowance of dew or hoar-frost. And the same process on the slopes would entitle the valley to the drainage of cold air from the higher levels; but the deposit of water-drops in the air which fills the valley seems to require loss of heat from air itself by radiation on its own account.

Fig. 11 *a–f*. Dappled skies.

a. Salcombe, summer, 1911. (Cave.)

b. Hampshire Downs, 1923, August 6. (Cave.)

c. Hampshire Downs, 1923, August 5. (Cave.)

d. Aberdeen, 1917, November 11. (Clarke.)

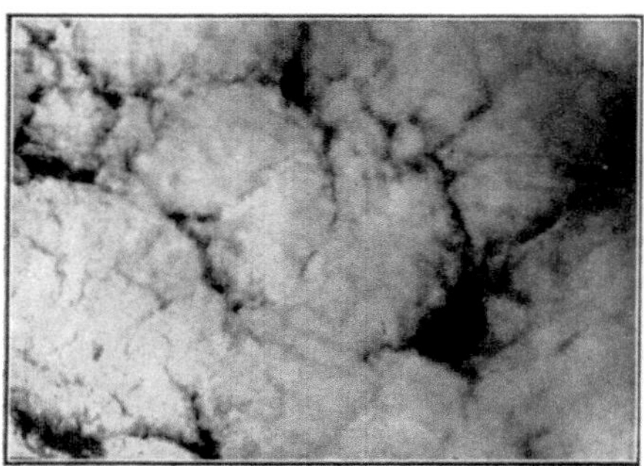

e. Lyme Regis, 1923 October 5. (Cave.)

f. Aberdeen, 1921, April 9. (Clarke.)

Dappled skies

Our next page of pictures shows something of what the pageant of weather can produce in the way of the picturesque. We begin (fig. 11*a*) with a mackerel sky from Salcombe made up of many sets of parallel stripes of very high cloud like the stripes on a mackerel. We call such clouds cirro-cumulus. They form heaps, and because they are so high they are regarded as made up of ice-particles not of water-drops. They are not thick enough to keep out the sun; they diminish the sunlight but do not prevent its throwing a shadow of objects on the ground.

Another example, fig. 11*b*, of cirro-cumulus, this time from the South Downs, looking north-west, shows an arrangement in small heaps and not in furrows. They form no shadows; what looks like shading in the picture is made by the blue of the sky seen between the little flocks of cloud.

Beneath the cirro-cumulus are two pictures of alto-cumulus, figs. 11 *c*, *d*, similar in form but lower in the sky formed by water-drops and thick enough to cast shadows on their neighbours; one was taken from the South Downs by Mr Cave looking south-east on 5 August 1923, and the other by Mr Clarke at Aberdeen on 11 November 1917.

Fig. 11*e* makes what may be called a "close-up" of the same kind of cloud directly overhead at Lyme Regis in October 1923, very like heads of cauliflower whatever that similarity may imply.

Fig. 11 *f* takes us back to Aberdeen and Mr Clarke for a specimen of one of the lower types of eddy-cloud, called strato-cumulus, which has abundance of shadow, and clouds the sun as it passes over. The picture shows the clouds advancing from seaward, a series of rough ridges with a bright bounding line of clear sky away out to sea.

Mechanical formations

The examples of the pageantry of cloud which we have exhibited are apparently the result of general conditions. They certainly will have derived their water from the surface, and in their form future generations may perhaps decipher their life-history at which at present we can only guess. But we can find some examples in which recognisable local influences have produced very striking effects.

Fig. 12 *a–e*. Mechanical Formations.

a. Mt. Pico, Azores. (Chaves.)

b. Hampshire Downs, 1924, April 21. (Cave.)

c. Zürichberg, 1907, August 16.
(de Quervain.)

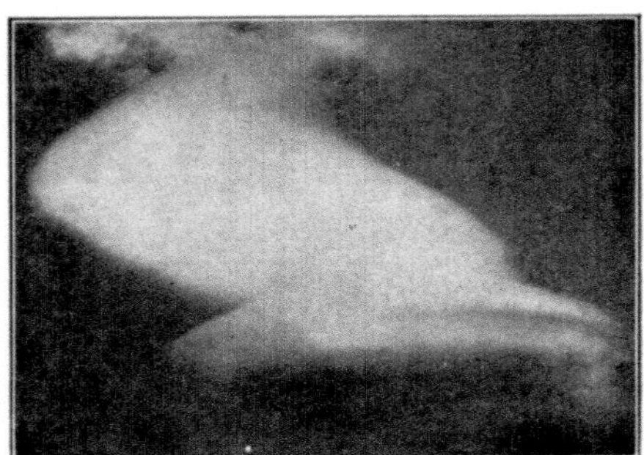

d. Etna, 1897, April 21. (Mascari.)

e. Invergordon, 1929, September 23.
(H.M.S. 'Furious', G. A. Williams, R.N.)

The most eloquent perhaps is fig. 12*a* with a stereo-photograph by the late Col. Chaves of the volcanic island of Pico in the Azores with a detached cloud so well known and so very like a whale as to have acquired in the locality the name Baleia, the Portuguese for a whale. It forms on the lee-side of the mountain. In this particular picture on the windward side is a sea-gull making its way upwind; the Baleia remains stationary.

Fig. 12*b* shows a cloud of the same type formed in the wind blowing from the east on the South Downs in April 1924. These clouds we call lenticular because they are shaped like a thick lens, from the same kind of analogy as the Portuguese Baleia.

They furnish a good example of the deception which nature may practise with the pageantry of her atmosphere. The wind actually blows through these lenticular clouds. It does not carry them away. The locality of the cloud remains the same but the material of the cloud is always changing. In this respect it is like a waterfall, seen from a distance it looks peaceful enough; the fall remains in the same place but the water does not, it goes through the fall. So the air goes through the cloud, with drops forming as it enters and dissolving as it leaves. The peaceful-looking cloud that caps a mountain practises the like deception. Let no one expect to find calm at the top when it is capped—quite otherwise—it is a place of bitter wind.

If airships had been invented earlier these clouds might easily have been called airship-clouds. Automatically the atmosphere makes them take the shape that airship-designers have thought the best for escaping friction. Fig. 12*c* from Switzerland shows a very obvious specimen of this type with a background of the ominous type of alto-cumulus castellatus, a specimen of which was exhibited in fig. 9*a*.

Finally there are two fantastic shapes, one fig. 12*d* the Contessa del Vento which forms in the wind-shadow of Mount Etna in circumstances similar to those which give Baleia behind Mount Pico, and a cloud somewhat similar, very mechanical in its appearance, from the hilly neighbourhood of Cromarty. But in this case as with the Contessa del Vento the features of the cloud suggest something of the nature of a whirl or vortex; the mechanism is probably more directly the result

Fig. 13 *a–c*. Haze.

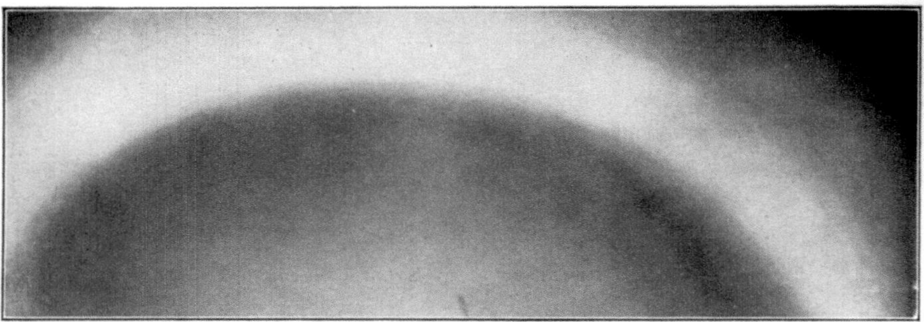

a. Halo at Aberdeen, 1910, September 30, 10h 05m. (Clarke.)

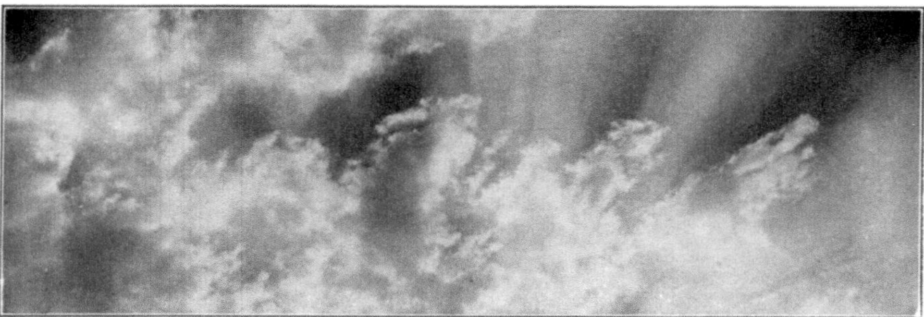

b. Aberdeen. (Clarke.)

c. Hampshire Downs, 1912, January 25. (Cave.)

of energy derived from the condensation of the water-vapour than in the other instances of mechanical formation when the cloud is little more than a visible index of the curved track of an air-current. Such whirls are perhaps more reasonably entitled to the epithet thermo-dynamical than simply mechanical.

Haze

Next we wish to illustrate the hazy cloud called nebula or simply haze. It may be found at various heights. One of the most common forms is found at great heights, 8 kilometres or more, and is called cirro-nebula. It is known to be composed of the very small ice-crystals, common to all clouds that carry the name of cirrus, because it often shows the bright rings slightly coloured which surround the sun or moon. The rings are at a distance of about 22° from the centre of light and are known as halo-rings. They can be explained by the behaviour of rays of light passing through ice-crystals; but water-drops with their spherical shape cannot give haloes—rainbows are their contribution to the pageant of the sky and sometimes coloured rings of corona close up to the luminary.

So cirro-nebula may be regarded as the highest of the nebular clouds, which are so faint as to have no form of their own and make themselves manifest only by the dilution of sunshine until the sky appears milky rather than blue. The philosopher finds in the halo undeniable evidence of the existence of ice-crystals in the air.

Underneath the halo-picture is a striking picture by Mr Clarke of clouds which remind us of the castellated cloudlets of figs. 9*a* and 12*c*, and are illuminated by the sun behind them which throws into prominence not only the cloudlets themselves but the nebula between them or above them or below them, it is hard to say which. These may be particles of water as are probably also those of the nebulous picture, fig. 13*c*, of Ditcham Park, Hampshire, on 25 January 1912 by Mr Cave which shows something of a cloudy nature but has not the definite boundary of the fog in the valley of fig. 10*c* or of the strato-cumulus of fig. 10*b*.

Fig. 14 *a–e*. Dust.

a. Snow-dust, Mount Everest. (Mt Everest Committee.)

b. Smoke dust in London. (Owens.)

c. 1920, October 11. Dust-horizon. (Judge.)

d. Dust-cloud in Iraq. (Lawson.)

e. Dust-whirls in India. (Baddeley.)

Dust

Finally we may give some illustrations of the dust that is found in the atmosphere and sometimes rivals water in its pageantry. The drops of water are actually formed in the atmosphere itself by the condensation of vapour; but there is no condensation in the case of dust. Yet it is found at all heights in the atmosphere. It is projected into the air on a magnificent scale by the eruptions of volcanoes and in a manner which cannot be called magnificent by ordinary chimneys; but it can also be lifted mechanically through the action of the wind on the loose dust at the surface by a process which is studied under the name of turbulence. To begin with we make a junction with the water-clouds by showing a picture of Mount Everest with a cloud formed by ice-dust blowing off the summit (fig. 14*a*). And then a corresponding picture, fig. 14*b*, of a cloud of smoke-particles blowing away from a chimney in London, quite as impressive in appearance.

Beneath these two, as fig. 14*c*, we show the curious appearance of false horizon formed by the top of the surface layer of dust carried into the air originally by the turbulence of the wind, but falling out again so slowly as to make it possible for the eye to distinguish something like a level surface in the case of fig. 14*c*, which is for 11 October 1920 from the Air Ministry. It is photographed by Mr A. W. Judge from 450 metres (1500 ft) above the apparent horizon which was itself at 1250 metres (4000 ft) above ground. The appearance of the veil of dust is made the more definite by little heap-clouds, cumulus, floating within the dusty air below the surface which is recognised as the dust-horizon.

The two pictures in the lowest line show a dust-storm in Iraq, fig. 14*d*, photographed by Flight-Lieut. J. F. Lawson, and some dust-whirls moving over sandy soil in the neighbourhood of Lahore from sketches by P. F. H. Baddeley published in 1860. The difference in the appearance of these two pictures is not without interest. The dust-storm would be attributed to the turbulent motion of the surface layer of air in a strong wind without any suggestion of how the wind originated. The other asks for some explanation of why the air-motion coiled itself into whirling columns when the general movement was not violent enough to raise the dust.

Fig. 15. Clouds seen from above.

a. France, 1918, August 15. (Douglas.)

d. Belgium. (Jaumotte.)

We will complete our cloud-pageantry by pictures of clouds from above which have been obtained in great numbers since flying came into vogue. Fig. 15*a* shows the upper surface at a level of 500 metres of the low cloud which we have called strato-cumulus. It is from France in August 1918 by C. K. M. Douglas. It shows more corrugation of the surface than the valley-fog. Two other pictures, figs. 15 *b* and *c*, by the same officer show more rugged upper surfaces, one of a higher cloud-layer classed as alto-stratus, and the other of a more vigorous strato-cumulus.

Fig. 15. Clouds seen from above.

b. France, 1918, August 15. (Douglas.)

c. France, 1918, August 25. (Douglas.)

e. Belgium. (Jaumotte.)

These are supplemented by two pairs of stereo-pictures figs. 15 *d*, *e* from the collection of M. Jaumotte, Director of the Belgian Meteorological Office, each pair presumably obtained by successive exposures for the same cloud, with a very short interval, from an aeroplane travelling horizontally. For the use of the lantern-slides from which the pictures are reproduced we are indebted to Sir George Simpson, Director of the Meteorological Office, Air Ministry. For stereoscopic views of clouds the distance apart of the two lenses is not the three inches of an ordinary stereoscopic camera but much greater.

THINGS THAT TRANSPIRE

Fig. 16. Waterspouts.

a. N. Pacific, 1916, August. (Mordue.)

Waterspouts, tornadoes, hurricanes and depressions

We may extend the range of our pictures of dust or water-drops in the air by three photographs (figs. 16 *a–c*) of the trunks of cloud in waterspouts by which the clouds in the sky over water are connected with the surface through a column or trunk of whirling air carrying an abundance of water-drops presumably with a core of very low pressure.

Close personal investigation of manifestations of that kind is generally avoided by cautious seamen, though some are bold enough to attempt their destruction by gunfire. How precisely the construction of the column is arranged we must leave for the present with the remark that it seems to be very like the most destructive form of atmospheric activity, namely the tornadoes of the so-called temperate lands (fig. 17), which are developed with moderate intensity in regions where cold wind invades a warm current and with especial violence in the lowlands of the United States, the valleys of the Mississippi and its tributaries, between the Rocky Mountains on the west and the coastal ranges on the east. Apparently the plains are an open road north and south by which water-laden air can be transported from the Caribbean sea to supply the region of the Great Lakes, and cold air can come down south from the Canadian plains. These conflicting currents develop an intensity of destructive activity that is only imperfectly achieved in other parts of the world. Perhaps the Euro-Asiatic

b. N. Pacific 1916, August. (Mordue.)

continent escapes that kind of visitation because its high land is arranged west and east instead of north and south.

In imagination it is possible to think of a series of stages exhibiting the same kind of behaviour ranging from the street-eddy which sometimes makes a visible whirl of dust and leaves, through the water-spout or tornado, very destructive but not very extensive, to the tropical hurricane of the West Indies or the typhoon of the China seas, which certainly represent whirls about calm centres, and by extension to the cyclonic depressions of temperate latitudes, the principal actors on the stage of the temperate zone, which display at least some of the charac-

Fig. 16 *c*. Lake Geneva, 1924, August 3. (Mercanton.)

Fig. 17. By favour of W. J. Humphreys. Four views of a tornado cloud at Quinn Ranch, Tree Canyon, five miles south-west of Gothenburg, Nebraska, on 24 June, 1930, 18h. (Mrs Roy Homer.)

From left to right:

1. The clouds before the onset of the tornado.
2. The funnel dropping, seen from a point one-half to three-quarters of a mile east.
3. Approaching Quinn Ranch.
4. The moment of striking the Ranch.

[Note: Gothenburg is on the Platte river near North Platte, 600 to 700 miles up the valley of Missouri and Platte river, on the 100 mile slope between the 1000 and 2000 ft contour, lat. 41° N.]

Ice-crystals.

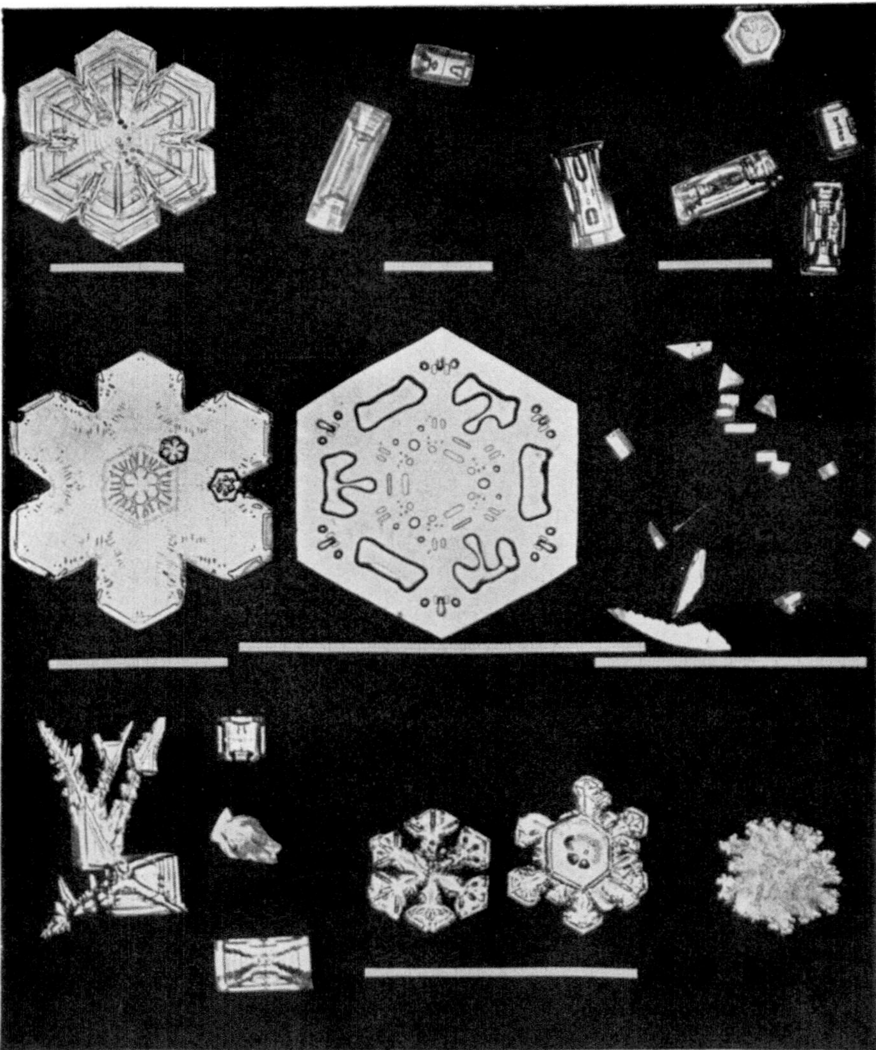

Fig. 18 *a*. Photographs of ice-crystals from the collection of W. A. Bentley. The white strips under the several photographs represent in each case a millimetre. (*Monthly Weather Review*, Washington, November 1924.)

teristics of the tornado but may be more than a thousand times as great in the area affected and cause less than a thousandth part of the material damage of a hurricane or a tornado.

Particles adrift

Let us supplement our picture-gallery of clouds by some illustrations of the material of which they are formed. There is little to show about water-drops which differ only in size. Only very small ones, of which it would take between 1000 and 6000 side by side to make an inch, are carried about as clouds. The larger ones, between a fifth and a five-hundredth of an inch in diameter, fall as rain unless they get broken up into fragments in consequence of being in too great a hurry to get down. With rain-drops we are all familiar.

Fig. 18 *b*. Stud-shaped snow-crystals. (J. M. Pernter.)

Ice-crystals on the other hand would require a whole gallery for an adequate display. They can aggregate themselves in beautiful patterns. The late Mr W. A. Bentley of Jericho, U.S.A., spent his life in collecting and photographing crystals that fell as snow, and obtained some forty thousand pictures. Some thousands of them are now included in a book recently brought out by Prof. W. J. Humphreys. The beauty and variety of the forms are in every way remarkable. Generally speaking they are thin plates and always show a structure based on a six-pointed star. A few are exhibited in fig. 18*a*, and there they are provided with white lines to show the dimensions of the crystals. A millimetre (the twenty-fifth of an inch) is a very liberal estimate of their size. Some specimens of ice-crystals which are not flat plates but have a structure like collar-studs are shown in fig. 18*b*.

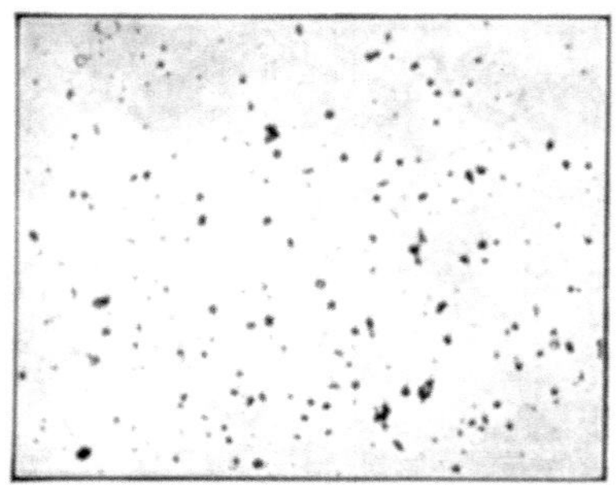

Fig. 19 *a*. Dust, London, 1922, March 26. (Owens.)

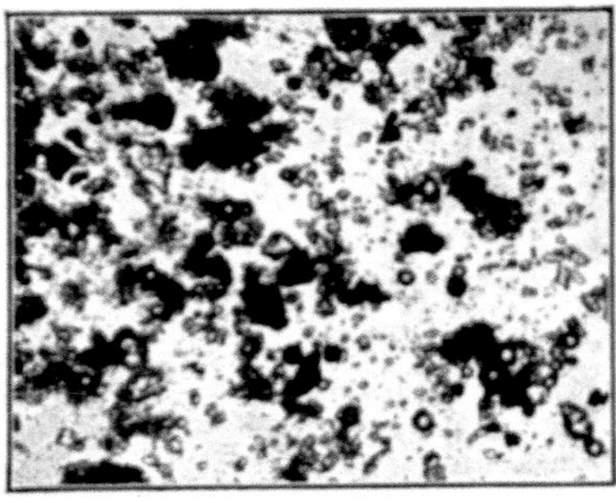

Fig. 19 *b*. Dust, Cheam, 1922, August 11. (Owens.)

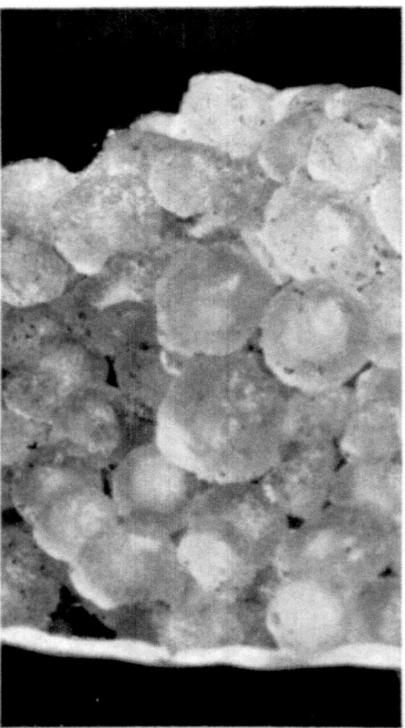

Fig. 20. A plateful of hail-stones about one-half natural size. (Royal Meteorological Society.)

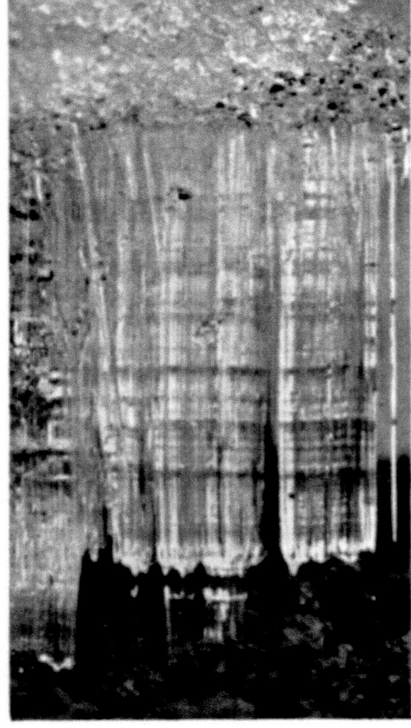

Fig. 21. Ice-columns formed on the top of the ground in Japan. (The Master of Sempill.)

Here we may be allowed to interpolate two figures (Fig. 19) to show particles which were taken from the air but are neither ice nor water. They have been brought under the microscope by Dr J. S. Owens for *The Smoke Problem of Great Cities*. The first picture illustrates dust in London air in March 1922 magnified a thousand times, the second a much more crude collection from haze at Cheam in August of the same year magnified five hundred times. In both cases the particles are very small compared even with ice-crystals, though they are comparable with the smallest cloud-drops.

And we must also remark that ice has a wonderful facility for displaying remarkable forms which are not included under the general term snow-crystals. The crystals which can associate themselves one with another to form snow-flakes can also spread out to form frost figures on window-panes and large masses of ice can be aggregated in the atmosphere as hail-stones (fig. 20). We may add too a curious columnar structure of ice formed at the surface of the ground in Japan (fig. 21), presumably by the freezing of successive supplies of water condensed from the ground underneath.

Another familiar form of ice-deposit is rime which collects on the projecting stems of trees in fog-laden air (fig. 22). The lower picture is a formation of ice on railings during an ice-storm. Both pictures come from Blue Hill Observatory.

We might pass on to exhibit also the aggregation of ice in glaciers and icebergs and various forms of sea-ice, there is, indeed, no limit to the possibilities of ice in the picturesque; but perhaps we have given enough to recommend the use of the camera to the traveller who is interested in the pageantry of nature.

Fig. 22. Rime in the garden and a coating of the rails in an ice-storm. (McAdie.)

Spring-frosts and other aspects of weather

Apart from the formation of ice, frost may appeal to us in many ways. We are familiar with the disastrous effects of ground-frosts in spring on growing potatoes and in the autumn on the flowering dahlias. Fig. 23 is a photograph taken in the Thames valley by W. H. Dines of a walnut tree that made a brave show of foliage early in the month of May. Ground-frosts at night are not infrequent in valleys in that month, and on one frosty night the lower foliage of the tree was so severely bitten that it withered and perished leaving the lower branches bare while the upper part of the tree above the layer of freezing air escaped the frost and continued to serve the tree's purpose of growing walnuts.

Fig. 23. Spring-frost. Benson, Oxon, 1922, May 31. (W. H. Dines.)

It would be possible to present a number of pictures which would illustrate the various aspects of weather that impress the observer and ask for explanation; but the possibilities are so numerous that the selection may be left to the discretion of the reader who has a camera and wishes to use it. As a supplement on page 106 we add for guidance an alphabetical index of weather that might be illustrated with the aid of a list of abbreviations drawn up originally by Admiral Beaufort for those who go down to the sea in ships and modified to meet some modern requirements.

Capital letters are sometimes used for intense examples of the weather represented ordinarily by small letters. Out of the 24 letters in the service of the Daily Weather Report it would be interesting to know which offers the best opportunities to the camera. It would naturally be pointed at something represented by a capital.

Fig. 24. Lightning flash and its likeness. (W. J. S. Lockyer.)

Electrical discharge

Passing from the ordinary photographs which have claimed our attention there are many other things that transpire. Some of them depend on the special behaviour of light. In the forefront we may place a picture of spontaneous illumination which is so familiar that it has a common name quite apart from any scientific implications, namely lightning. Known throughout the ages and associated with thunder it was recognised as electrical by Benjamin Franklin.

The mind is apt to rest in a verbal explanation like that; but it is not quite an end of the story. Side by side with the lightning flash from a photograph by W. J. S. Lockyer is a copy of the River Amazon traced by him from a school atlas. The close analogy must have some meaning and the likeness of both to the branching of a tree must raise the question whether the main trunk of the lightning is in itself the reservoir or does it connect two branched systems, one visible and the other invisible, as a tree-trunk connects the visible branches in the air with the invisible roots in the ground?

THE TRICKERY OF LIGHT AND COLOUR

In this story of the pageantry of weather we have been concerned with what an interested spectator can see, and have used as illustrations some examples of what the camera has recorded. The camera and its pictures are themselves the result of the experimental work of physicists and chemists, and as we think about them the question at once arises whether what we can see and what the camera records are the same things.

To find an answer we require some knowledge of what the eye can see and what the camera can record. With certain limitations, to which brief reference will be made later on, the camera is a trustworthy guide to what the eye can infer concerning the positions, shapes and motions of clouds and other things. An obvious failure of the picture reproduced by the camera is the lack of the colour in which the eye rejoices. The camera's picture shows only gradation of tone from light to dark, whereas for the eye the sky is blue, the trees are green, the clouds are sometimes white, often grey, sometimes red, sometimes what we call black; sometimes they show patches of rainbow colour; and if we would deal with the question of the relation of the camera's record to the eye's experience we must understand what colour means for the eye and what seeing means.

The physicists tell us, and by their experiments would demonstrate to us that all the colours, and indeed light and darkness themselves, depend upon the rhythmic titillation of nerves at the back of the eye by waves which travel from any object which is seen, through the intervening transparent medium, the air, and through the transparent lens in the front of the eye. On the analogy of the travelling waves of the sea they tell us about the length and frequency of the waves that produce this sensation of light. They say that they all travel across open space with the amazing speed of 186,000 miles a second. Whether the experiments would explain to us exactly what it is that is here at one moment and 186,000 miles away at the next tick of the clock we hardly like to say. While the rate at which they travel is so extraordinarily large, the length of successive waves in the travelling train from crest to crest is extraordinarily small, and is different for different colours.

The longest wave-length of light which an ordinary eye can see, that of the deepest red, is measured as less than the thousandth part of a millimetre, the twenty-five thousandth part of an inch, and the shortest, a very deep violet, has a length of about half that of the deep red. Between those extremes are waves which affect the eye as different colours of which seven are identified by name from violet the shortest, through indigo, blue, green, yellow, orange to red the longest.

Travelling through space the shortest and the longest keep together and the whole bundle coming from the sun produces an impression which we call white. But travelling through the transparent substances with which we are familiar, air or water or glass, the red lags behind the violet and making use of this in suitable apparatus we can show all the seven colours arranged side by side in a band of all the colours of the rainbow which the physicists call a spectrum. The transition from one colour to another is gradual not by finite steps. The instrument that exhibits the spectrum is a spectroscope. A triangular prism is one of its parts and prismatic colours are another name for the dispersion of colour which is exhibited in the spectrum of white light.

With the wave-length so small and the velocity of travel so great an enormous number of waves have to pass into the eye in a very short time, about 1000,000,000,000,000 per second for the violet and half that number for the red.

Physicists tell us further that for other purposes, including wireless telegraphy, nature provides a whole series of waves that are exactly similar to waves of light but have wave-lengths too long for us to see; and there are also waves too small for us to see—ultra-violet—which are, however, very effective in the camera's picture and are also highly esteemed by doctors as a valuable characteristic of sunlight coming through clear air.

We must go a stage further. An object may be visible because it is itself luminous like the sun or a fire or a candle or a lamp, all exhibiting high temperature which accounts for the luminosity; but the number of self-luminous objects is limited not by the failure of the object to emit waves but by the incapacity of our eyes for seeing the waves which the colder objects emit and which are too long for us to see.

An object can be made visible by the echo of light from the sun which is redistributed by reflexion from the illuminated surface. The moon is visible not by its own light but by the light received from the sun and redistributed by the moon's surface. So all objects are visible in the sun's rays or in the rays of a lamp if they have any capacity for redistributing the light. Those that have no capacity of that kind must always look black; those like clouds that have the capacity for redistributing everything that falls upon them regardless of colour look white, so long as the light which illuminates them is itself white and the cloud has something darker than itself for a background.

One other property we must refer to. We have mentioned a transparent medium like air, water or glass as transmitting the waves that enable us to see the object from which the waves originated, but the transmission is not always perfect, indeed we are nearer the truth in saying that it is never really perfect, it transmits some wave-lengths better than others. Coloured glass is the best kind of example. Green glass for example transmits what appeals to the eye as green, red glass as red, blue glass as blue, but disposes otherwise of the other wave-lengths. The intricacies of the subject are endless. Let us mention some examples.

Sunlight passing through the smoke of a wood fire makes the smoke blue for anyone looking across the sun's rays and catching the light that is redistributed by the smoke, but the sun itself seen *through* the smoke looks red. So the sun's rays crossing the atmosphere make the sky look blue but the sun seen about sunset through the lower layers of the atmosphere looks red, deep red if the atmosphere is heavily charged with dust or smoke, and the sunshine thus reddened when it illuminates a cloud on the other side of the sky, which ought to be white, makes the cloud look red. So we must distinguish between a red object which will look red in sunlight which carries all the colours that go to make up white, and an object like a red cloud which looks red because red is the colour of the light by which it is illuminated. In green light it would look green and in white light white.

Let us say a word about black and white, words which are often used about clouds. For the physicist colours correspond definitely and strictly with wave-length; but for the eye, depending on the sensations excited by the waves, the correspondence is not definitely with wave-lengths; colours will mix to give other colours in the eye, blue and yellow mix to make green, red and green to make yellow, and quite a lot of mixtures to make white. Full black means that the object is not taking any part at all in the illumination that reaches the eye. A white object can only take part in the illumination if it gets light from some luminous source that it can redistribute as an echo. So for lack of illumination on its visible side a white object will look dark if there is a bright background for it, and will look white if it gets a little illumination and has a dark background.

In considering the different shades or degrees of illumination something ought to be said about the useful grey. A mixture of black powder with white powder is grey, so is a cloud of black spots on white paper; darker or lighter means more black or less black. Any one of our cloud-pictures will disclose that fact to the reader with a magnifying glass. Grey is often used about the sky not as an indication of colour but simply of the degree of illumination. Black is no colour at all, white on any occasion a special combination of many colours, and grey the degree of illumination of one part of the sky compared with its surroundings.

So we get the extraordinary complication of a sky with a variety of clouds. A cloud in sunshine with the sun behind the spectator looks brilliantly white, but in the shadow of other clouds with the bright sky as a background it looks dark, sometimes so dark as to be called a black cloud. Even smoke that we are accustomed to call black may look what the eye would call white if it is provided with a background of darkness.

Thin white clouds let enough light through them to illuminate clouds in their shadow; but clouds so thick that very little sunlight can penetrate to the under surface will be dark, almost as dark as night.

Nothing perhaps in the sky offers so much opportunity for thinking out details as the illumination and the colour of the clouds that float in the air.

Fig. 25. Colour vision.

(*a*) Ultra-violet. (*b*) Infra-red.
San José from Mt Hamilton, 1924. November 9. (Wright.)

Colour and visibility

The camera itself is by no means insensitive to colour. The difference of its behaviour towards light of different wave-lengths is shown by the following examples.

An ordinary photograph can only express black or white by the chemical effect of the light on the photographic plate, but with suitable colour-screens for the lens the light which passes through can be restricted to special colours, red or violet. When a plate, made sensitive to deep red, is exposed with a suitable screen we get a picture in which red has blackened the negative and will show white in the print; the other colours, being intercepted by the screen, produce no effect.

Fig. 25 shows two photographs of the same view, one made by ultra-violet light which will pass through a screen of metallic silver, and the other by the extreme of red. The farther distance in the picture by red is much clearer and better defined than the other, and if we agree that the intervening particles are too reflective of ultra-violet light we must conclude that the red light is not so much reflected and we get a clearer picture. So distant views are clear when we limit our attention to what is red, whereas they are fogged when the view is not so limited.

Mirage

While concerning ourselves with the intricacies of light and colour we have been thinking of the light from the sun or some other luminary as transmitted in straight lines; but that is only true for transmission through free space. We have allowed the reader to suppose that if he pointed his camera at a distant object he would find the object straight ahead in the line of sight.

That is not always true; the optical line is not always a straight line. In consequence of the variation of density in the lower layers of the air the light sent out from a distant object is always bent and sometimes actually reflected; the object is distorted and may even appear inverted or the visible image duplicated.

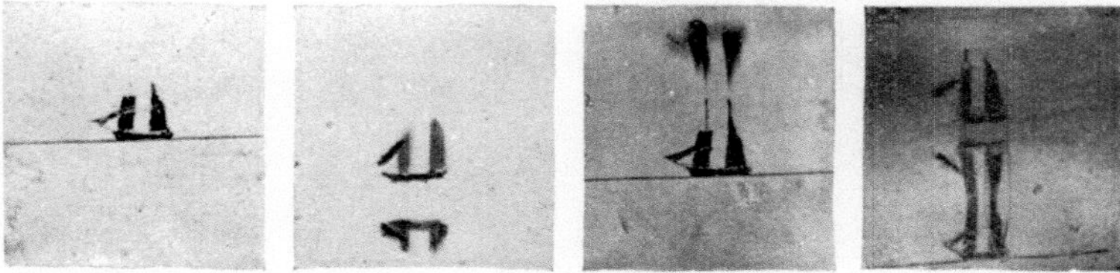

Fig. 26. Photographs of the shape of a ship through a cell of water and syrup

Such phenomena, included in the title mirage, are specially frequent in the regions where the surface air is exceptionally hot or exceptionally cold. On the Euphrates during the War the seeing was so bad that an engagement had to be postponed.

Fig. 26 illustrates the mirage in a very effective manner. It is not, however, a mirage but the collection of images of the outline of a ship marked on glass and photographed by A. Mallock through a solution of sugar.

Fig. 27. Seen at Ramsgate on 1 August 1798 by Prof. S. Vince. (*Phil. Trans.* 1799.)

Similar pictures are actually seen; an example (fig. 27) was seen by Vince and Scoresby at Ramsgate on 1 August 1798.

Fig. 28 *a*. The pageant in Hampshire. 1931, July 21. (Cave.)

THE WIND'S PART IN THE PAGEANTRY OF WEATHER

After the interlude showing some of the things that the camera can display, let us return to our first presentation of the drama of the clouds by two pictures which suggest more definitely the important rôle of the winds, one (fig. 28*a*) from the south of England by Mr Cave of July 1931, and the other (fig. 28*b*) from Japan by Count Abe.

The clouds are obviously borne along and therefore associate themselves with the story of the winds that carry them. The currents of the upper air hold the key to the action of the drama, but for the story to be realised by the unsuspecting student, the currents that carry the clouds must be nursed by their environment with a delicacy of attention

Fig. 28 *b*. The pageant in Japan. Fujiyama and its clouds. (Abe.)

that the best nursemaid cannot rival. Let their path slope downward ever so little and the clouds are gone—vanished into vapour; set them going upward ever so little and they become fractious or even violent and may express their resentment by adding thunder and lightning to their rain.

In the Count's picture the clouds in the immediate neighbourhood of the mountain are obviously of the mechanical kind that we have called "lenticular". The wind blows through them where they are formed; but in Mr Cave's picture the winds were carrying the clouds, bringing them, in fact, towards the camera, and the same may be true of the more distant cloud-bank in the picture of Fujiyama.

Another example of the wind's part in the pageantry of weather comes from Commander E. R. G. Baker showing the witches' cauldron at Gibraltar under the influence of a levanter, a notable east wind of the

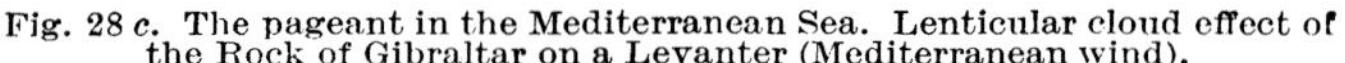

Fig. 28 c. The pageant in the Mediterranean Sea. Lenticular cloud effect of the Rock of Gibraltar on a Levanter (Mediterranean wind).

Mediterranean. The photograph is by Leading Seaman Hayward and it is included by permission of the Topical Press Agency, Ltd.

It is the winds of the upper air that carry clouds, and at the outset we have to distinguish between them and the winds that ripple the water, raise the dust or move the trees. Straws show the way the wind blows, so does a weathercock; but they are not the best guides to the way the clouds go. If a weathercock happens to be included in the observer's field of view in an ordinary urban or woodland situation, he will be interested to note that the weathercock which obeys the surface-wind points to the left of the direction from which the clouds are coming, generally about four "points" of the compass, 45°, to the left. That is the ordinary deviation of the surface-wind over the land from the line of motion of the lowest clouds. But that is only part of the complication. Clouds at different levels can often be seen moving in different directions.

The drift of clouds from the south-west over a wind from the north-east was a familiar experience of the Greeks who accused the north-easter of greediness, because he drew even the clouds towards himself whereas the others helped to speed them on their way. That feature of the behaviour of a north-easter is just as common in many other parts of the world to-day as it was in Greece more than two thousand years ago and is still. It means that the clouds that are visible are carried from the south-west by a wind from that direction above the surface wind from the north-east.

Every picture of the sky is a scene in the pageant of the atmosphere, an episode in the play of the wind and weather of the world which the human race has been trying to understand for thousands of years. The most impressive presentation of the drama is a new cloud-atlas which has been engaging the attention of the meteorological authorities for some years and is now published.[1]

THE SUN'S PART

The first step towards understanding the play is to realise that the whole creation is dependent upon the heat and light which come from the sun by a process which is called "radiation", a name suggested by the sun's radii, or in English rays; by a similar process when the sun is gone the supply is dissipated from the earth and cloud by radiation which is invisible to our eyes.

The radiation conveyed in sunshine keeps the earth supplied with what physicists call energy—heat is one of its forms, motion another, electric current between the two ends of the filament of a lamp another. It is "energy" that we pay for in electric lighting or heating at a certain price for a B.T.U., Board of Trade unit, a kilowatt-hour. When all the king's horses and all the king's men have done their best we shall get electricity all over the country at something like a penny for a unit of energy, and meanwhile at that price whenever the sun shines on a square

[1] *International atlas of clouds and of states of the sky.* Office National Météorologique, Paris, 1932. Recently supplemented by a Spanish edition of the Atlas, by *Cloud around Mt. Fuji* by Count Abe and an Atlas for the Philippine Islands by Father Deppermann S. J. of Manila.

yard, making allowance for slope and the murkiness of the atmosphere, we get the equivalent of a pennyworth of electricity every hour—an open space 50 yards square would get ten pounds' worth, a square mile of sunshine is worth ten thousand pounds an hour.

And on the other hand when there is a clear sky at night every square yard of earth would be throwing away something near a farthing's worth of energy every hour.

This process of the transfer of energy by radiation is one of the great mysteries of nature. At the beginning of the nineteenth century it was agreed that the transmission was by means of waves across the ether of space. But since agreement was reached about the transmission of radiation which we appreciate as light, many other things have been brought under the same rules: the radiation of heat from a fire and the radiation from an electric spark that introduced us to wireless, and now all the experiences of the wireless world as well as the world of radium and all its tribe.

Sunshine and parasols

The whole story is long and complicated, but here is a little side-show which we will borrow from the story of the action of the play and which will help us to keep in mind something of the part that sunshine plays in the drama, and of the beautiful intricacy of the pageantry of floating clouds, such clouds as we see in fig. 29.

The air, as we know, is made up mainly of the gases oxygen and nitrogen with some water-vapour which is variable in amount because it may be increased by evaporation from any water-surface or diminished by deposition as dew or cloud or rain or snow. Now when the sun shines on the air, asking for a passage for its energy, it is allowed by the nitrogen and oxygen but not quite so easily by the water-vapour. As long as the air is what we call transparent the invisible water-vapour appropriates part of the energy that would go through if permitted, and stores it as increased warmth in the mass. Any dust or smoke that is mixed up in the air will help the water-vapour to plunder the travelling energy.

So when there are isolated patches of moist air in sunshine there will be robbery in every one, and each one will become warmer than its

Fig. 29 *a–c*. Sunshine and parasols.

a. Hampshire Downs, 1923, October 15. (Cave.)

b. Dorpat, 1927, August 21. (Kousnetzoff.)

c. Java, from Mount Tjikorai. (Braak.)

more innocent neighbours. But when the parcels of air get warmer than their neighbours, their neighbours push them up out of the way, so the worst thieves are all pushed upwards, expanding as they go. And, as we shall see, expanding means cooling; so in the end the plunderers may find themselves no warmer than they were, and if the environment is suitable may after all be cooler and so seriously cooler that each one of them will develop into a cloud.

Then the story begins: the sun shines on the cloud as it does on the transparent air; but now the water-drops become mirrors, millions of millions of tiny spherical mirrors which, as we know, do not give the sunshine a chance of getting through, but not for covetousness; they reflect the sunshine and become themselves magnificently illuminated, the beauties of the sky. Three-quarters of the sun's energy is rejected and dissipated, and there is less water-vapour to steal the energy that passes the drops. So the process of stealing practically stops as long as the drops are there to reject, or, as the scientific say, to reflect the sun's rays. Each cloud uses its water-drops as a sunshade; let the sun shine as it may the clouds go on their ways like ladies with their parasols.

Think of that when next you look at a sky dappled with clouds that float in the sunshine as in some of the pictures of this chapter, and if you make it out, think about the way fog forms in the air and disappears.

To this pretty story some realists may object that the amount of energy absorbed from the sun's rays by water-vapour in the atmosphere is so small that it is not entitled to the star part for which we have cast it.

The point thus raised is an interesting one which we cannot deal with fully in the present chapter. Here we may note that the classical experiments on the absorption of radiation by water-vapour, while exhibiting its great power of absorbing long waves, which we cannot see, allow also that it may absorb a little group of short waves within the range of the sun's light and heat.

Perhaps it is enough for our purpose, perhaps not; direct measurement of solar radiation must decide, and meanwhile there is good authority for putting upon the stage things "just so strange that though they never did they might happen".

THE ORGANISED STUDY OF THE PLAY

Now that we have some insight into the pageantry of the sky, the prevailing influence of wind and sunshine representing motion and radiation, two of the rudimentary forms of physical energy, and have had before us, as an example, an interpretation of the action in one small aspect of the drama, we can make a brief and general survey of the whole field.

Throughout the ages mankind has tried to interpret for his own advantage the weather-drama of which the visible clouds present one scene.

Modern weather study, the science of meteorology, implies an organised system for tracing the development of the play over a much larger field and for deciphering the physical processes which connect successive pictures as cause and effect. There is in fact no limit to the range of the meteorologist's vision short of the atmosphere of the whole globe.

Progress has been slow. Those who are now continuing the effort are the inheritors of the achievements of a long series of workers, philosophers, physicians, poets and historians whose names are familiar words in the household of the science. To the ancient Babylonians we owe the measurement of time and the ordering of the seasons according to the position of the sun in its annual path through the twelve signs of the zodiac; and to the ancient Greeks our geometry, the measurement of the earth, in daily use. As a science meteorology begins with Hippocrates, the physician, who about 400 years before Christ first expressed appreciation of the influence of climate. Speculation as to the physical causes of atmospheric changes begins for us with Aristotle, the Greek philosopher of the fourth century B.C., whose treatise *Meteorologica* became the text-book of physical science for the scholastics of the Middle Ages and the Renaissance. In the title were included all the phenomena which are observed in earth or air or sky. Since Aristotle's time many branches of "physics", the study of nature, have become separate sciences, and the word meteor on which the original meteorology depended has come to mean for most people only the shooting stars that roam through space, blaze up as they pass through our atmosphere and sometimes dig their own graves in the earth. The atmospheric disturbances caused by one of them on 30 June 1908 have become an event in meteorological history. Large craters attributed to falling meteors have been identified in Siberia, in Arizona and in central Australia.

Aristotle's pupil Theophrastus provided a careful study of the winds of Greece and gave us too a collection of weather-signs, a practical and proverbial side of the science, which we now call weather-lore, and which finds a place in the poetry of Hesiod before 700 B.C. It is resumed by Aratus the poet-physician of the third century to whom St Paul referred in his address to the Athenians. Virgil gave a Latin version in the *Georgics* in the first century, perhaps the most often quoted of all books on weather-lore. References to weather and its influence are abundant in classical literature and the historian Pliny, the elder, who sacrificed his life to his curiosity in the great eruption of Vesuvius in A.D. 70 records a good deal of natural knowledge of airs, waters and places.

With the fall of the Roman Empire scientific study languished until the seventeenth century when a new era was inaugurated by the practice of experimental work. To that century belong Galileo Galilei, inventor of the thermometer, Torricelli, his scientific assistant, of the barometer. Blaise Pascal, religious enthusiast and ascetic in France, ascertained the variation of pressure with height, Robert Boyle, philosopher and religious enthusiast in Oxford, established the relation between the volume of a parcel of air and its pressure. William Dampier, seaman, wrote a celebrated treatise on the winds of the globe, which were mapped by Edmund Halley, magnetician and astronomer. And from that we pass on in the next century to George Hadley, lawyer, who introduced the effect of the rotation of the earth on the winds, Joseph Black, physician, who introduced the idea of latent heat of liquids and vapours, and Horace Bénédict de Saussure of Geneva, geologist and physicist, who devised the hair-hygrometer for measuring the humidity of the air. Pierre Simon de Laplace, "the Newton of France", professor of mathematics in the military academy, set out the law of the variation of pressure with height.

And in the nineteenth century we find Charles Wells, physician, who gave the theory of dew. Then we pass on to John Dalton, schoolmaster, who gave the laws of the pressure of vapours as well as an Atomic Theory, Luke Howard, manufacturing chemist, who classified clouds, Louis Joseph Gay Lussac, chemist, who gave the law of expansion of gases with increase of temperature, Sir Francis Beaufort, Admiral, who provided a scale for wind and letters for weather, James Pollard Espy, teacher of classics, an apostle of the theory of storms, Elias Loomis, mathematician, who traced the travel of storms, H. W. Dove of Berlin, professor of physics, who wrote a celebrated work on storms, M. F. Maury of the United States navy, who advocated the development of the meteorology of the sea, Henry Piddington, seaman, curator of a museum, who invented the name cyclone, Francis Galton, traveller, enthusiast for and patron of science, who added anticyclone and devised weather-maps, U. J. J. Le Verrier, astronomer, and Robert FitzRoy, Admiral, who were the first officially to collect information by electric telegraph, Alexander Buchan, schoolmaster, who traced cyclonic depressions across the Atlantic, Buys Ballot, professor of physics, who formulated the relation between pressure-distribution and wind at sea-level, Hermann von Helmholtz, Army doctor, then physiologist, then physicist and then mathematician, who gave us the principles of atmospheric stratification and of atmospheric waves, Julius Hann, professor of meteorology, in Vienna, an encyclopaedist of his subject and author of the best known books on climatology and meteorology.

We ought not to forget the inventors of photography and transfer-lithography, without which meteorology would be badly impoverished; but meteorology is not alone in that position.

There are two sides to the usefulness of systematic notes about weather—one, which is classed as climatology, traces the rhythm of the sequence of weather at the place of observation and its irregularities; and arranges the result in a form which permits easy and effective comparison with other places, tells us in fact why some places are habitable and others not. The other, which is more strictly classed as meteorology, deals with the intimate physical relations of the whole sequence of events one to another.

On either side of the science but particularly on the physical side, everything depends upon the co-ordination of observations made all over the world. John Ruskin, a celebrated authority on art and other things in the nineteenth century, expressed very forcibly the peculiar character of the science in this respect in the transactions of a Meteorological Society formed in London in 1823 with the idea of a world-wide organisation in mind.

> The meteorologist is impotent if alone; his observations are useless; for they are made upon a point, while the speculations to be derived from them must be on space....The Meteorological Society, therefore, has been formed, not for a city, nor for a kingdom, but for the world. It wishes to be the central point, the moving power, of a vast machine, and it feels that unless it can be this, it must be powerless; if it cannot do all it can do nothing.

In spite of Ruskin's ambitious sturdiness, the Society found the task beyond its powers. Progress towards achievement had to wait until the electric telegraph made it possible to combine on a map within an hour observations from a sufficient number of stations to make the idea of a general view of the play of wind and weather possible.

To understand a situation means first of all to portray it, to get in mind the interaction of all the parts. In intimate connexion with this vital aspect of the subject we find a suggestion by the celebrated natural philosopher Clerk Maxwell, whose centenary was celebrated in 1931, that if you wish to understand the forces which control the motion of a body the best method of approach is to learn exactly what the motion of the body is. That was the method of approach in the case of the moon and planets which led to the idea of universal gravitation. So if we would make out the forces acting upon the atmosphere we must ascertain the motion of its parts, an enterprise which implies nothing less and nothing more than realising the drama and forming a pageant to represent it.

It is a task that has never been achieved and weather will never be understood until it has been; but many separate parts of the stage have been scrutinised and the information used for practical purposes.

When British universities include the study of weather as a subject for their students, a part of the practical examination must be to invite the candidate to look at the bit of the drama of the sky framed by a window, to say what he sees and what it means. It would test the examiner as well as the candidate. And there is perhaps no subject more suitable for illustrating what is meant by science, the gradual progress from the chaos of the first impression, through co-ordination and classification to order, and finally to the laws which even now are not attained in anything like the same degree as in astronomy. The drama of which one sees a very small section through the window is the same drama as that which is intended to be represented for a million times the area in a weather-map of Western Europe, or for a hundred million times in a weather-map of the northern hemisphere, such as is now issued daily from the Meteorological Office, and the test of success of the exponents of the science is the degree in which the drama is effectively represented on the map.

CHAPTER I

IDEAS OF THE DRAMA ANCIENT AND MODERN

Accuracy of narration is not very common, and there are few so rigidly philosophical as not to represent as perpetual what is only frequent or as constant what is really casual.

SAMUEL JOHNSON, *Tour of the Hebrides*

THREE BASIC IDEAS

Our story naturally regards weather-maps and the broadcast report of inferences drawn from them as displaying our conception of the dénouement of the drama, the latest stages in the scientific application of the study of weather. Other conceptions of the action of the play may be inferred from the records of the past. The information that emerges makes it clear that mankind has always felt the need of information about the future, and as history develops we can recognise the ideas about the interpretation of the drama of weather and the attitudes towards it that were favoured by the wisdom of the times. That is suggestive of an effort to meet a public demand, by real or assumed knowledge of weather and its control, rather than the orderly development of a science or of the material of which a science is built.

Three basic conceptions make themselves apparent; first, especially among the Greeks and the Norsemen, the personification of the elements and the control of the visitations as the personal concern of major or minor deities who might express tempestuous resentment by some form of atmospheric violence, thunder or lightning or wind, and who might be appeased by some suitable expression of sacrifice or submission.

Secondly, especially among the Hebrews, gradually approached by the Greeks, the idea of the sequence of nature as ordered by a benevolent and paternal deity to follow an appointed course that might perhaps be altered by adequate supplication in special circumstances but which for the most part conferred its benefits upon the just and the unjust alike.

Thirdly, in association with either, there is the recognition of relations between consecutive natural events which finds expression as "weather-lore". Therein we have a form of proverbial philosophy which could be appreciated equally by those who attributed favourable weather to the personal benevolence of a deity and by the natural philosophers whose primary concern is to co-ordinate events and seek an explanation by expressing the relationship as a natural law. Such laws are now regarded as inductions from the facts observed, not necessarily as cause and effect but as a sequence attested by experience.

Prediction is the natural exploitation of the knowledge implied by associations recognised or assumed, a mental deduction from accepted principles.

The controllers of the nations seem always to have had official advisers as to future policy and immediate action, and the weather finds its place naturally among the vicissitudes which have to be foreseen when any enterprise is under consideration. It was a serious responsibility to decide whether circumstances were "auspicious"; for the Romans the flight of birds or the entrails of sacrifices supplied the necessary information and so, in a way, were substitutes for our weather-maps. The "popular" attitude had necessarily to be attuned to the wisdom of the wise, which developed as a craft or mystery associated with religious practice as a guide to conduct, and subject to all the dangers of the unrestricted control of oracular information. All sorts of devices were adopted as contributing to the formation of expert opinion.

In a cursory survey it is not possible to deal separately with the basic ideas which we have summarised. They overlap to such an extent that separation is impracticable.

THE MEDITERRANEAN REGION AND ITS PECULIARITIES

Our information is derived mainly from the records of the countries surrounding the Mediterranean sea and the course of evolution was to some extent dependent on the natural conditions of those regions.

It is remarkable that the Mediterranean countries have no word for what we call weather, or perhaps the remarkable feature is that the Teutonic and Norse languages have such a word. In the old records of the Mediterranean region there are "times" with many qualifying adjectives, and seasons (a word derived from *satio*, a sowing) but no general word for weather as inclusive of any atmospheric phenomena.

In the Authorised Version of the English Bible the word "season" occurs sixty times, and "time" meaning season practically as many; but "weather" only four times. Two of them are in the striking reference to Hebrew weather-lore in our Lord's reply to a request for a sign from heaven. "When it is evening ye say it will be fair weather for the sky is red, and in the morning it will be foul weather to-day for the sky is red and lowering", and even in those two the word has been introduced in the translation in place of the words in the Greek text εὐδία, the favour of Zeus (ἐῦ Διός) for fair weather and χειμών, winter, or tempest, for foul weather.

In the Revised Version in one of the four examples (Job xxxvii, 22) "*golden splendour*" is substituted for "*fair weather*"; and in the fourth example (Prov. xxv, 20) where confidence in an unfaithful person is said to be "as one taketh off a garment in cold weather and as vinegar upon nitre", obviously "time" would be almost as appropriate. In all, the vital part of the description is the adjective, a "fair time", "foul time" or a "cold time" would do almost as well. A separate word for weather is indeed a literary luxury but a very present help for a meteorologist.

Words for separate phenomena of weather are not lacking; whirlwind appears many times in the Old Testament—it "came out of the south". Thunder, lightning, storm and tempest, rain, rain-showers, snow, hail are sufficiently abundant to indicate the common experience.

The Greeks also were well supplied.

"When ye see a cloud rise out of the west (δυσμή, setting sun), straightway ye say, There cometh a shower (ὄμβρος, a storm of rain); and so it is. And when ye see the south wind (νότος) blow, ye say, There will be heat (καύσων, scorching heat); and it cometh to pass. Ye hypocrites, ye can discern the face of the sky and of the earth; but how is it that ye do not discern this time (καιρός, season)?" Luke xii, 54–6.

Their word for hurricane, ὁ τυφώς, fits in with Arabic and Chinese to describe what we call a typhoon. Indeed, on account perhaps of its wealth of suitable words, ancient Greek supplies so many technical terms for modern meteorology—as for all other modern sciences—that a page in explanation of the general plan has seemed a necessary sequel to this chapter.

In these days, for weather the French use *temps*, the Italians *tempo*, the Spanish *tiempo*, all obviously derived from *tempus*. The number of words which we derive from that root is remarkable. Here are some: tense in the grammatical sense, temporal, temporary, temperance a characteristic of temper, temperament, temperature and tempest. The last four are peculiarly appropriate to the weather of the Mediterranean. Its main characteristics are seasonal, a very dry summer intervening between the cereal harvest in May and the fruit harvest in September, with occasional thunderstorms accompanied by violent winds. It was "the weary season" with the Greeks, the dog days, the season of quarrels; as it is with us to this day the season for the outbreak of wars, strikes and lock-outs. And then a winter with capricious rainfall between October and April associated with sudden and violent gales, so that "tempestas" characterises the temporary disturbances of the regular progress of the seasons and is easily regarded as indicating temporary disturbances of the temper of the controlling deities, and changes of temperature are the most notable features of the weather.

In more northern latitudes the changes are more gradual and more amenable to description as the ordered sequence of legitimate phenomena. The reasons for this peculiarity of the literary treatment of the weather of the Mediterranean as compared with more northern countries are in themselves an interesting chapter in meteorology of which we can only indicate some notable features. The region lies between the thirtieth and the forty-fifth parallels of latitude. The corresponding latitudes over the neighbouring Atlantic Ocean include what are called the "horse latitudes". The peculiar name is derived from the fact that persistent calms there were fatal to the transport of horses in the days of sailing ships. It is a region of permanent high pressure and calm seas with westerly winds on the north side and easterly winds on the south side fed by the permanent north-east trade-winds along the north-west African coast.

The Mediterranean is an inland sea, a protrusion of the horse latitudes eastward into the great land area of three continents. It is connected with the ocean by the narrow Strait of Gibraltar guarded by the Pillars of Hercules. On the east side beyond the range of the winter rains are the desert regions of Mesopotamia and Trans-Jordania. On the north are the well-watered lands of Europe, south of the Alps, and the mountainous extensions to the Caucasus with the Black Sea intervening. On the south side, scantily watered, are the Atlas mountains and the plains of Cyrenaica with the great North African desert behind them.

In summer the water of the Mediterranean is cool compared with the land to north or south and the conditions are not very dissimilar from those of the horse latitudes outside. An etesian or summer wind flows across the eastern region and travels down the valley of the Red Sea to help the supply of the Indian monsoon as a sort of counterpart to the drainage of air in the west to feed the trade wind. In the winter the water is relatively warm and the land cold, sometimes very cold. In winter and summer alike the sea takes all the drainage of the water from the high ground on both sides and in winter it continues the same relation with the cold air of the high land. Streams of cold air coming intermittently from north or east pour off the land-areas as the mistral of southern France and the bora of the Adriatic and the Aegean. On a smaller but often more violent scale they furnish an analogy with the flow of polar air from the north into the westerly current which is a sort of birthright of the latitudes between 30° and 50°. These invasions develop sudden and fierce cyclonic depressions that are carried by the westerly current along the sea from west to east sometimes penetrating to Iraq and northern India, while another part of the same cold drainage passes round the Atlas and helps the winter supply of the trade-wind. The Mediterranean countries call the winds by special names: etesian, gregale, scirocco, khamsin, bora, chile, tramontana, euroclydon, levanter, maestro. The only example of special name in Britain is the helm-wind of the English Lake district. We compact all our winds into the single word weather.

These are in brief the conditions which the references to weather in ancient literature have in mind. We must remember the possibility of a change of climate in the Mediterranean region within the period covered by the available records. Geologists tell us that in times gone by large parts of the world which are now comfortably habitable were covered with snow and ice, and they suggest that the last of these glacial periods was 18,000 years ago. What the cause of the glaciation may be is still a subject of discussion, but it is an accepted principle that a number of such glacial periods have occurred with intervening pluvial or rainy periods and dry periods.

The abundance and splendour indicated by the ruins of Palmyra, in an oasis of the Syrian desert between Damascus and the Euphrates, which was a flourishing capital in the third century A.D., is a suggestion of a change in climate due to failure of rainfall. On the other hand it is claimed that the natural produce of the countries of the Mediterranean region in crops and fruits is the same as it was two or three thousand years ago.

Changes in rainfall are said to be indicated in the variation of the rings of growth of the great conifers of California and elsewhere, and the whole subject of the variation of climate is a most attractive side-issue for the student of weather. Here we will only remark that the gods have other ways of expressing their anger besides tempests of weather, one of which, practically ineluctable, is the persistent covering of a region with sand until it is buried. Sand forms by the disintegration of rock and is carried by the wind and deposited wherever protection is available. And as the accumulation increases, the unfortunate people suffer without any remedy.

"Tryphon made ready all his horsemen to come that night; but there came a very great snow, by reason whereof he came not."

Two things are unconquerable in the long run, sand and snow, and sand is a sort of snow that never melts.

And on the other hand in countries where the falls of rain are relatively few and relatively heavy a rainy visitation in a hilly country is quite likely to wash away the soil from the uplands unless it is protected by the growth of trees. The United States furnishes some notable examples.

So, without careful inquiry, economic disaster is not a matter of legitimate complaint against the caprice of the weather-gods.

It is not surprising that the information about weather that has come down to us from the past is concerned mainly with what we may call the violence of weather. The regular rhythm of the season might call for little remark, but the temporary interruptions—tempests, whirlwinds and especially thunder and lightning—had to be accounted for; they provide the most impressive signs of their approach and leave undeniable evidence of their power. They are very suggestive of the behaviour of a temperamental personality.

For the Greeks, Zeus was in charge of the air, Poseidon of the sea and Pluto of the solid earth with its minerals. So in the Homeric age in the tenth century B.C. a flash of lightning was Zeus's method of putting an end to an attack of the Greeks upon the Trojans. According to the historian Livy the consul Valerius attacking an enemy's camp about 470 B.C. discontinued the attack on account of a sudden storm of thunder and hail, which would now be called a line-squall, soon followed by calm and bright weather. "He felt it would be an act of wilful impiety to attack a second time a camp defended by some divine power." And even the poet Horace in the Augustan age, the first century B.C., acknowledged himself convinced of the personal interposition of Jupiter, the Latin equivalent of Zeus, when a thunderstorm broke out of a clear sky.

So the regular course of the seasons and fair weather (εὐδία) might be attributed to the benevolence of Zeus and the interruptions of that tranquillity ascribed to his personal intervention.

The perplexity of an intelligent mind in face of natural events can hardly be better expressed than in the statement of the problem of weather in the Book of Job, somewhere in the period of the Greek philosophers. It finds some sort of echo in the meteorology of the Book of Enoch, a compilation of religious writings belonging to the end of the second century B.C. The wording of the Authorised Version is itself rather bewildering—we quote from the Revised Version.

> Elihu also proceeded, and said, Suffer me a little, and I will shew thee; for I have yet somewhat to say on God's behalf.
>
> * * * * * * *
>
> Behold, God is great, and we know him not; the number of his years is unsearchable.
>
> For he draweth up the drops of water, which distil in rain from his vapour: which the skies pour down and drop upon man abundantly. Yea, can any understand the spreadings of the clouds, the thunderings of his pavilion?
>
> Hearken ye unto the noise of his voice, and the sound that goeth out of his mouth. He sendeth it forth under the whole heaven, and his lightning unto the ends of the earth.
>
> After it a voice roareth; he thundereth with the voice of his majesty:

Great things doeth he, which we cannot comprehend. For he saith to the snow, Fall thou on the earth; likewise to the shower of rain, and to the showers of his mighty rain.

Out of the chamber of the south cometh the storm: and cold out of the north. By the breath of God ice is given: and the breadth of the waters is straitened. Yea, he ladeth the thick cloud with moisture; he spreadeth abroad the cloud of his lightning: and it is turned round about by his guidance, that they may do whatsoever he commandeth them upon the face of the habitable world.

Dost thou know the balancings of the clouds, the wondrous works of him which is perfect in knowledge? How thy garments are warm, when the earth is still by reason of the south wind?

* * * * * * *

And now men see not the light which is bright in the skies: but the wind passeth, and cleanseth them. Out of the north cometh golden splendour: God hath upon him terrible majesty.

(The Book of Job xxxvi, 1, 26, etc.)

Job's answers in the examination are not recorded but they could not have got many marks for there follows: "The Lord answered Job out of the whirlwind, and said, Who is this that darkeneth counsel by words without knowledge?...Where wast thou when I laid the foundations of the earth?" And the same question has been suggested by the whirlwind throughout the ages of the history of meteorology and still reaches us from the typhoon of the China seas and the tornado of the American plains. Time after time "words without knowledge" have darkened the counsel of the weather-wise, and even now we are not quite out of their shadow.

THE GRADUAL ACCUMULATION OF MATERIAL

It is a matter of some interest to trace the gradual transition from the idea of personal influence to the recognition of what we call natural law. It began very early. Herodotus, the Greek historian, in the fifth century B.C. tells us that the crew of a ship in distress brought the storm to an end after three days by sacrificing victims to the Nereids (the daughters of Nereus and Doris who were in charge of the Mediterranean) and other powers, and he adds laconically "or perhaps it ceased of itself". And Aristophanes, a satirical dramatist of the same century, in his comedy of the *Clouds*, stages a wrangle between a sceptical philosopher Socrates and an unsophisticated countryman Strepsiades as to whether thunder was Zeus's doing or the natural result of a clash of clouds in a vortex of air.

The philosophic or scientific method of approaching the subject is by observation and experiment to trace the physical processes which connect the phenomena of to-day with the conditions of yesterday, and so lead on to a prospect for to-morrow; but the task is not by any means an easy one and the first stage of the process—the co-ordination or classification of the facts observed—may be provisionally expressed as maxim or proverb while waiting for the necessary experimental work. The weather-signs of Aratus, Theophrastus and Virgil which are referred to in the Prologue must have represented common knowledge at the time, and with various additions they have been handed down from generation to generation and from country to country, and appear as a formidable array of 200 pages in a book of *Weather-Lore* by Richard Inwards with a bibliography of 160 items. And all the time almanacs have been giving predictions of weather for a year ahead based ostensibly on inferences drawn from the positions of the planets and stars.

Climate as a subject of study appears in the fifth century B.C. in a work on *Airs, Waters and Places* by Hippocrates, the "father of medicine". Aristotle's great treatise on physical science, *Meteorologica,* regained for Europe, became the ultimate authority on the subject for the Continent in the Middle Ages, and held its place until the eighteenth century. All things in the earth and sky were included in the survey. It discussed the formation of rain, dew, rainbows, rivers and other phenomena. And as one reads the text, conscious of the coherence in natural science introduced by the recognition of the laws of conservation of mass and of energy which found expression in the nineteenth century, one can hardly fail to wonder whether some other laws with a similar grip of facts will add a like precision to modern ideas current in the highest physical circles about the relations of space, time, radiation and electricity.

Judged by our standard the most pertinent observations in the *Meteorologica* are those which relate to the classification of the winds of Greece. The idea of personification was still retained. The winds had names according to their character. Boreas is the rude north wind, Zephyrus the mild west wind, Lips the homeward leading south-wester, Notus the rainy south wind, and Kaikias the greedy north-easter.

Fig. 30. The Tower of the Winds. SW face. Skiron in front.

Since Aristotle wrote his treatise many of the subjects which he included in the *Meteorologica* have established themselves as separate physical sciences which can be developed by individual effort, and the name now connotes the remaining phenomena of the atmosphere which, as Ruskin rightly explains, can only be studied effectively by the cooperative effort of students in all countries. The misnomer of the old name is sometimes remedied by calling the surviving science aerology or aerography.

The winds are the subject of a dissertation of Aristotle's pupil Theophrastus, whose *Book of Signs* has already been mentioned. The winds were still classified by their personal characters; the allocation of direction is not by the measurement of what we call orientation from north or south but by the rising or setting of the sun in summer and in winter or the supposed locality of origin.

We find the beginning of our modern method of orientation in the Tower of the Winds in Athens (fig. 30) which housed a water-clock and showed by sundials the time of year as well as the time of day.

It is octagonal in plan and at the top of each face is a sculptured figure of one of the winds with its name. All the figures are winged, have suggestive clothing and carry symbolic gifts, thus associating science with the idea of personification. The wings of the wind are characteristic of the earlier conception of the weather-drama.

Here perhaps we may offer a question for those who are interested in the study of weather. Why did the constructor of the Tower of the Winds picture the winds as flying from left to right? Why in fact is

Fig. 31. The figures of the Tower of the Winds arranged in clockwise order, reversing the circulation suggested by the original arrangement.

Notus the south wind, shown emptying a jar of water, and pursuing Eurus the uncertain south-easter, while Lips the south-wester is shown to follow the rainy Notus, to be followed in turn by Zephyrus the west wind, the birthright of the region, and then by the north-wester Skiron (Thraskias or Argestes), an old man holding an inverted firepot suggestive of cold winds, squalls and thunder-showers at the end of a depression; and so on all round the tower: a circus with the centre on the left of its line of procession?

That is the order which we moderns, on the initiative of Dove's *Law of Storms*, cannot fail to recognise as associated with the passage of a stormy cyclonic depression along the Mediterranean, whereas arranged the other way round the figures would have been suggestive of tranquillity and fine weather which we associate with an anticyclone, not nearly so interesting for the inquiring mind. By way of illustration we have provided in fig. 31 a diversion of the original arrangement which exhibits the figures in anticyclonic order. It gives an indication of what is meant by personification. The three names in the front row read backwards.

So far as the Greeks were concerned, physical measurements were restricted to lengths, angles and weight, which were sufficient for the astronomy that deals with the reckoning of time, and for establishing the cycle of the seasons. Perhaps for that reason practically nothing was added to the science after Aristotle until the development of experimental methods in the seventeenth century, beginning with the invention of the thermometer and barometer.

WEATHER-LORE

In the meantime the imperative necessity of some knowledge of weather found expression on the one hand in astrological prediction and in the proverbial philosophy of weather-lore on the other.

A good deal of weather-lore relates to what we may call the raggedness of the rhythm of the seasons. The course of the sun across the sky for any part of the world has been established by the astronomers as perfectly rhythmic with extraordinary accuracy, and if mundane things were so arranged as to repeat conditions with the same accuracy as the repetition of the sun's track across the sky, summer and winter, rainfall, sunshine and warmth would be just *da capo* from the beginning of each year the same as the one before. But that is far from being the case. All sorts of irregularities intrude themselves and the various experiences become proverbial; and so we get cold weather, known as the borrowing days, and a blackthorn winter in April, ice-saints in May, the tradition of St Swithin for July and August, dog days, the days of the dog-star, a very ancient tradition of heat in August, St Luke's summer for warm weather in October and St Martin's summer for similar conditions in November. These are the things which interested folk in days gone by and are even now the incidents about which newspapers are curious.

The association of the proverbial interruptions of the rhythm with certain phases of the calendar finds its development in modern meteorology by curves of mean daily temperature. With that as guide Alexander

Buchan recognised certain cold periods in the temperature-rhythms of Scotland, and they have reached the public through the press as Buchan's cold spells which he says "occur from year to year with very rare exceptions" and according to his investigation are determined and regulated by the winds.

The association with the course of the sun is reasonable enough and perhaps it is natural if not so reasonable to associate the deviation from the rhythm with the phases of the moon, the influence of which is so easily recognised in the tidal fluctuation of the sea. We may remember that before the lighting of our streets and the daily paper which carries the date, the phases of the moon afforded a convenient method of counting time-intervals between the day and the year. Experience would naturally be associated with the moon's phases in a manner somewhat similar to the association of the larger features with the seasons of the year. So it may have come about that the moon has been held to be responsible for weather whatever it happens to be. Fifty years ago, or even up to the introduction of weather into broadcasting, the meteorology of the educated and of the uneducated alike was the expression of associations with the moon and its phases which even now has not been forgotten. There is still left a seaman's proverb that the wind changes with the tide.

There is an easy transition from the staging of notable occurrences as phases of the year or the month to the use of notable occurrences as prognostics of the future. One weird tradition associates impending rain or fair weather with the moon on its back. It can hardly be justified, for the way the moon appears to lie depends merely on the relative position of the sun, and goes through its regular sequence with the year.

Other proverbial prognostics may be regarded as the vague expression of experience; readers may have their own: here are some:

Wet Friday, wet Sunday; Rain before seven fine before eleven; Rain from the east twenty-four hours at least; Fine weather coming if there is enough blue sky to make a sailor's breeches; A red sky at night is the shepherd's delight; A rainbow in the morning is the shepherd's warning; If the sun goes pale to bed, 'twill rain to-morrow so 'tis said; The north wind doth blow and we shall have snow.

As examples of others with less justification we may note the contradictory proverbs of the ash before the oak or the oak before the ash, and a full crop of berries means a hard winter.

There is a vast literature of weather-lore, some of it like the Shepherd of Banbury's rules based ostensibly on experience, some a collection of proverbs and traditions as in Mr Inwards' volume, some of it astrological or like the last-quoted example based upon the habit of human psychology which regards any and every occurrence as a sign of the future whereas the scientific habit would claim that they should be regarded as consequences of the past. And even that limitation does not prevent the peculiarities of our psychology from intruding upon our science. Every novel phase of human activity comes under suspicion of affecting the weather; the relation of gunfire to weather has a history of its own in southern Europe where periodically there is a demand for means of bombarding thunderclouds as a protection of the vineyards against hail. In northern Europe gunfire is regarded as a rainy influence; and wireless has been accused of responsibility for rainy weather.

WEATHER WISDOM AND PREDICTION. ASTROLOGY

In weather-lore we find the beginnings of the practice of using the observations of natural phenomena to afford some indication of weather to be expected. It goes back a long way and again the impressive influence of thunder takes a leading place.

From the cuneiform records of Nineveh in Assyria in the time of Assurbanipal, the seventh century B.C., we are told "When it thunders on the day of the moon's disappearance the crops will prosper and the market will be steady". And the Chaldeans, neighbours of the Assyrians, developed a scheme of prediction about weather and fertility deduced from the occurrence of thunder in each of the twelve months, that is to say between successive disappearances of the moon. The practice was continued in mediaeval almanacs as thunder-signs, *signa tonitrui*, but has lost its vogue.

The most ambitious effort at prediction came from the Babylonians and Chaldeans who were distinguished as astronomers and mathematicians. They associated the phases of human experience with the relative positions of the planets, "wandering stars", and thus originated what has been called "astrology", the science of the stars.

That aspect of the subject of weather-prediction merits separate consideration, because from its beginning with the Chaldean astronomers it has continued through the ages as a sort of separate science using nominally the same methods of development as those of the orthodox physical sciences, the same precision of observations; but like alchemy and the search for the elixir of life it was practised in a cloud of mystery by special experts quite outside the atmosphere of transparent candour enjoyed by the philosophy of the seventeenth century. In that way it is naturally associated with other oracular crafts and mysteries; with witchcraft and again with certain forms of priestcraft.

It passed from Asia to the west and though strongly opposed by the authorities of ancient Rome it became dominant in Europe during the Middle Ages as a method of prediction of the course of human affairs. Its exponents had a world-wide reputation; Nostradamus in the sixteenth century, Franciscus Allaeus (Yves de Paris) and William Lilly in the seventeenth.

Astrology was regarded as effective in matters of health and disease on the ground of a perfect correspondence between the several parts of the human body as a microcosm and their counterparts in the external universe as a macrocosm. In weather-science as a means of prediction it went so far as the regular issue of almanacs with the salient features of weather throughout the year as auxiliary information for the calendar with its astronomical features and its religious festivals.

It has continued to the present day. It still has its advocates and its regular journals. *Old Moore's Almanac*, amongst others, with predictions by Zadkiel or Raphael, have a circulation far beyond that of any orthodox meteorological work except the broadcast of to-morrow's weather. They give the tempestuous features of the weather of the year in advance as well as the fortunes of peoples, states and rulers. The awkwardness of having the country's future set out in advance was so clearly realised in England that the issue of almanacs was made subject to the censorship of the Archbishop of Canterbury.

The practice of astrology is still confined to experts. Its phraseology is cautious; only the barest statements are given, in very technical language, of the astronomical facts or the inferences which are based upon them.

Abundant traces of its influence are still to be found in the languages of the present day. Lucky stars are still predominant for favoured persons. Ascendancy is an astrological term, so is disaster, and so indeed is influence itself. It was a part of the social organisation.

If we are surprised that imperfectly reasoned inferences are accepted by so large a fraction of the community we must remember that popular judgment depends much more on striking instances than on any reasoned statistical examination of the facts. In the case of weather-lore a few examples of the change of weather with the moon when change is very much desired are sufficient to carry conviction. The outstanding successes and perhaps also the outstanding failures live in the memory, while the results of unnumbered similar efforts for less important occasions are forgotten.

Nobody computes the statistical results of observations of the favour of a black cat or the protection of a mascot on a motor-car, while the predictions by Lilly of the Great Plague and Fire of London go down to history with his almanac.

Indeed, the difference between the past ages and the present would appear to be more conspicuous in language than in mentality. Personification is still practised. In continental languages the elements of weather, and other inanimate objects, may be masculine or feminine though our English does not recognise the distinction.

For us personification is limited to ships or engines. But we have our own habits of creating personalities. It is not unfair to say that following Newton the eighteenth century personified force, and to a certain extent worshipped the force of gravity which was past finding out; the modern physicist seems to put the velocity of light in the same category. In the nineteenth century we personified the cyclone and the anticyclone; an Australian meteorologist gave names to those which crossed his path. We are still left with that inheritance; though, so far, the twentieth century seems disposed to personify, instead, the air-masses whose motion makes the cyclonic depression, and our present aim is to induce the reader to look upon the sequence of weather as a drama having its own dramatis personae. Progress in the science of meteorology is really dependent upon the proper choice of objects for personification. One of the favourites has been the mean values which are useful mainly for comparing with other means or with departures therefrom. A good deal of scientific energy has been and is being sacrificed on the altar of mean values.

To anyone who watches the clouds from where he lies it seems quite likely that we are personifying the clouds as travellers and perhaps more than they deserve. We tabulate their motion and plot the direction on charts. In Count Abe's picture and some others in the Prologue we agree that some of their representatives are travelling, others not. It requires a careful expert to avoid a mistake in the relation of the motion of the air to that of the cloud.

And again some of us are still at work on the scientific side of the pursuit of the relation between mundane affairs and the heavenly bodies, begun by the Chaldeans. Though the study was somewhat debased by charlatans, who succeeded them and spoiled the reputation of astrology, inquiry is not exhausted.

Strenuous efforts are still made to represent the general conditions of weather as controlled by the wandering stars in their courses, intermediately through their influence upon the radiation of energy from the sun.

By their combined action depending upon the changes in their relative positions they are thought to produce the sunspots which we can see and count, and other disturbances of the sun's activity; processes which are quite in keeping with the recognised canon of physical science.

These influences are supposed to find expression in the regular recurrence at fixed periods determined by the motion of the planets, of weather-conditions, rainy, warm or cold. Sunspots have an irregular periodicity of about eleven years which has its counterpart in mundane affairs, for example in the level of Lake Victoria in equatorial Africa, and may be due to the combined gravitational influence of Jupiter and Saturn. And a period of 35 years in weather has been on the scientific mind for centuries and lately elaborated as Brückner's cycle.

Efforts are made in various directions to use this kind of information as a basis for the prediction of weather for a long time ahead, just as eclipses are predicted by astronomers. The sailing might perhaps be plainer if the earth were all land or all water or all ice, or if the listener could somehow take in the experience of a group of years at one sitting instead of riveting his attention on the one he is experiencing. It is part of the weather-drama; we have something to say about it in a subsequent chapter. The play as presented by the weather-wise will be set out.

METEOROLOGY AND MEDICINE

As an example of the less ambitious prediction based on weather-lore, which would include the barometer as a weather-glass with its legends of rain, much rain, stormy on one side, change, fair, set fair and very dry on the other, we may quote some verses which are attributed to Dr Erasmus Darwin, the grandfather of Charles Darwin, a celebrated physician of Lichfield and the leading spirit of the Lunar Society, a coterie of eminent scientific folk in the English Midlands at the end of the eighteenth century. In a way the verses summarise within forty lines the weather-lore of Theophrastus, Aratus and Virgil to which we have already referred.

The hollow winds begin to blow,
The clouds look black, the glass is low,
The soot falls down, the spaniels sleep,
And spiders from their cobwebs peep.
Last night the sun went pale to bed,
The moon in halos hid her head.
The boding shepherd heaves a sigh,
For, see! a rainbow spans the sky.
The walls are damp, the ditches smell,
Closed is the pink-eyed pimpernel.
Hark! how the chairs and tables crack,
Old Betty's joints are on the rack;
Her corns with shooting pains torment her
And to her bed untimely send her.
Loud quack the ducks, the peacocks cry,
The distant hills are looking nigh.
How restless are the snorting swine!
The busy flies disturb the kine.
Low o'er the grass the swallow wings,
The cricket, too, how sharp he sings!
Puss on the hearth, with velvet paws
Sits wiping o'er her whiskered jaws.
Through the clear stream the fishes rise
And nimbly catch th' incautious flies.
The glow-worms, numerous and bright,
Illumed the dewy dell last night.
At dusk the squalid toad was seen
Hopping and crawling o'er the green.
The whirling dust the wind obeys,
And in the rapid eddy plays.
The frog has changed his yellow vest
And in a russet coat is dressed,
Though June, the air is cold and still,
The mellow blackbird's voice is shrill,
My dog, so altered is his taste,
Quits mutton bones on grass to feast.
And, see yon rooks, how odd their flight,
They imitate the gliding kite,
And seem precipitate to fall,
As if they felt the piercing ball—
'Twill surely rain—I see with sorrow
Our jaunt must be put off to-morrow.

There is perhaps a touch of irony in the verses for, altogether, thirty-one symptoms are enumerated in order to justify the simple prognosis "it will surely rain". Any one, presumably, might be sufficient, and we are left to conjecture how many of them must be in evidence before the patient would be justified in acting upon the conclusion.

Instead of Dr Darwin, Dr Edward Jenner, the introducer of vaccination, is sometimes named as the author of the verses. In either case it is a distinguished physician who gives us the copious but careful diagnosis of the actual condition of the weather and the prognosis of coming change; and that leads us to remark upon the close analogy that there is between the science of meteorology and that of medicine.

Going back to rudimentary civilisation we find weather-practice associated with magic and witchcraft. The weather-man was at the same time the medicine-man. In both departments of knowledge all human beings are concerned. While medicine has to try to understand the inner working of the living body, meteorology has to deal with the environment in which the body lives and moves and which itself seems sometimes to have a close analogy to a living organism. Scientifically both have to depend upon "diagnosis" of conditions or symptoms and their meaning, and "prognosis" of the course of events; and unfortunately throughout the ages both have been liable to the treatment of the subjects by those who in the case of medicine are classed as quacks and in meteorology as cranks.

In this sense it is not at all unreasonable of the astrologers to regard the human body as a microcosm and the close environment of the atmosphere which is a part of its life as a macrocosm with a definite correspondence between the several parts of each. With the barometer the meteorologist lays his fingers on the pulse of the atmosphere; the thermometer tells him about its temperature which is ultimately responsible for any feverish activity of the weather. The anemometer that tells him about the circulation of air is his stethoscope. His substitutes for the physician's examination of skin and tongue are his observations of clouds. The energy supplied to our bodies in food is measured in what are called calories; so is that supplied to the atmosphere by the sun's rays. The processes of digestion have their counterpart, but here the analogy becomes a little too intimate.

In fact it is perhaps not quite reasonable to regard our bodies as separate from the environment in which they live. If we were to say that human personality is nothing more than a nucleus of mentality in a material environment which we call the atmosphere we should have to spend more pages than we have at our disposal in order to controvert the statement.

Every phase of the behaviour of the atmosphere recalls some characteristic of personality either of a man or a woman, sometimes of both. A physician's attitude towards his patient is not so very different from that of a forecaster towards the creature that is represented in his weather-map.

With surgery, of course, it is different, though there are people in the world who have tried a kind of surgery on the atmosphere, and would like to try again. They cost more but are not nearly so successful as their prototypes.

What there is of science in weather-lore as in medicine must depend upon our understanding of the natural or physical processes which are the real expression of the relation of cause and effect. And in both, the physical processes are still the subject of persistent investigation not yet by any means complete. In the meantime in the case of both the sciences the action which we adopt in any combination of circumstances must be guided by such knowledge as we possess, and in so far as that knowledge is rational the sciences render their service to humanity. We may supplement our excursion into history therefore by some notes of the part which the knowledge of weather plays in the service of the world.

THE STRESS OF PUBLIC SERVICE

From the rough outline of the course of development of the study of weather it will be clear that throughout the ages mankind has been demanding for its guidance more information than science has been able to give, just as in the science of medicine human needs have always asked for more assistance than the profession had at its disposal. It is not so with all the sciences. Astronomy, for example, is so efficient that by its aid a ship can ascertain its own position anywhere on the sea;

with a little magnetism to help, it can start from any port in the world and find its way to any other port as directly as if it were on a line of rails, provided that the weather does not interfere by fog or hurricane. The measurement of time has been carried to such a degree of perfection that by the aid of wireless and its own chronometers a ship may know beforehand, wherever it may happen to be, the exact time of sunrise, noon and sunset.

Towards perfection of that kind meteorologists may perhaps look forward; but at present the cautious confine themselves to a "further outlook" which occasionally reaches a few days ahead; they have to decline the demands of the daily press to know what kind of winter will follow a wet summer? will fog be unusually frequent? what are the prospects for the summer holidays? or the Christmas week? and a hundred other questions that have been asked every year in the past and will be asked every year in the future.

The stress of service has hampered the progress of the science. Apart from the opportunity which it gives to what may be called rash speculation or imposture, it places the science in an awkward position. It is the habit of scientific folk who work in the seclusion of a laboratory to draw inferences from their experience and arrange experiments to test them; but to publish their inferences before they have been tested by experiment is, to say the least, unusual. It is not done in the best scientific circles and it provokes remarks about meteorology not being an exact science, meaning that its predictions (like those of any other science) are not always correct. Yet that sort of premature announcement is what the forecaster has to make; the stress of necessity overrides the laws of conduct. The mathematician has his equations, the physicist his formulae; but the forecaster has only a file of maps to illustrate the situation with some provisional general principles and his own experience to guide him. He cannot himself arrange the tests, he has to accept what the atmospheric conditions display. In a properly regulated world he might have continued the practice adopted by the Meteorological Committee of the Royal Society from 1867 to 1879 and confined his attention to accounting for the recorded sequence of events. If required to broadcast, he might have explained why it rained yesterday instead of stating whether it would rain to-morrow, and the explanation would not have lacked interest to listeners.

That is, in effect, what the practice of forecasting used to be before the development of wireless telegraphy. The forecasts were indeed drawn up and made known to the public in the newspapers; but very few saw the morning forecasts until the late evening, or the evening forecast until the next morning, and then only the town dwellers who saw the papers and could range themselves with the forecaster in his comparison of prospect with reality which was an official duty; but now twelve million listeners can, if they please, learn the forecast within an hour of its issue, and the attitude of the listener is determined by his temperament rather than by that of the forecaster.

With forecasting, as with other dramatic efforts,

> A jest's prosperity lies in the ear
> Of him that hears it, never in the tongue
> Of him that makes it.

THE PRACTICAL ACHIEVEMENTS OF THE STUDY OF WEATHER

Apart from the prediction of to-morrow's weather, meteorology has many achievements to its credit. Real progress has been made with the project which Ruskin sketched so boldly to the Meteorological Society of London in 1839.

For the land the exchange of information between the nations has provided us with a knowledge of the climate of every habitable part of the globe, with the exception of some parts of northern Africa and Central Asia, and has provided a certain amount of information about what may be expected in the uninhabited regions of the Arctic or Antarctic. We shall exhibit some of the results in a later chapter by means of diagrams which are more eloquent than words. We still lack necessary knowledge of rainfall over the sea in order to complete our survey of the main characteristics of the behaviour of air-currents.

The study of weather, originally the work of private persons, has become a part of government. By international agreement in 1853 the maritime nations have been in co-operation to obtain systematic information about weather at sea, and starting from a conference in 1872 an international organisation has been formed, of which the directors of the meteorological services of all countries are members, for the exchange of information about weather from every habitable part of the globe, and also for the distribution of information by wire or wireless from stations on land and ships at sea, so that the official meteorologists of any country can obtain within about an hour of the time of observation material to form a map of the distribution of weather within, say, 1000 miles of the locality. For the northern hemisphere the scheme has been developed so far that in Britain in the course of every twenty-four hours eight thousand facts about the weather are placed at the disposal of the forecaster who acts as utility officer; about half of the eight thousand facts are home grown, and half imported from abroad; and the organisation is so efficient that within three hours of the time of the morning observation a map is made and forecasts drawn up and distributed to shipping within wireless range and to the millions of listeners on land if they care to listen, and in little more than an hour printed copies of the map and inferences are available for distribution by post.

Quite a considerable number of the facts placed at the disposal of the meteorological service are derived from observations of the upper air, with the natural aid of clouds and the special investigation by means of the drift of pilot balloons and observations in aeroplanes.

Originally begun in the nineteenth century by the travellers in free balloons, since 1896 the investigation has been developed on an international scale. So now with the aid of geography we can present a picture of the results of the investigation extending up to 100 kilometres to show the Kennelly-Heaviside layer that helps the wireless signals to go round the earth instead of flying off into space, a high temperature layer that makes it possible to hear sounds at a hundred miles or more away, the tropopause at a height of five to ten miles that separates the troposphere, with which we are all personally familiar and which shows temperature diminishing with height, from the stratosphere which shows little or no variation of temperature with height and is only penetrated occasionally by energetic airmen or people like Professor Piccard. Some details are represented in fig. 40.

It is these developments of the science which are most expressive of its utility to-day and it is the modern hero, the airman, that most feels the need. It is not too much to say that a knowledge of the play of wind and weather is a matter of life and death for the airman. In all forms of aeronautics, ballooning, airship navigation, aeroplaning or gliding, the possibilities of weather are the first consideration and the experience of yesterday helps out the knowledge of to-day.

What the knowledge of weather is now for the airman it was for the seaman when the navigation of the seas was young. In the early days the use made of the knowledge was to stay ashore when there was any suspicion that the weather gods were likely to be unkind.

For fifty days after the turning of the sun, when harvest hath come to an end, the weary season,[1] sailing is seasonable for men. Thou shalt not break thy ship, nor shall the sea destroy thy crew, save only if Poseidon the Shaker of the Earth or Zeus the King of the Immortals be wholly minded to destroy. For with them is the issue alike of good and evil. Then are the breezes easy to judge and the sea is harmless. Then trust thou in the winds, and with soul untroubled launch the swift ship in the sea, and well bestow therein all thy cargo. And haste with all speed to return home again; neither await the new wine and autumn rain, and winter's onset and the dread blasts of the South Wind, which, coming with the heavy autumn rain of Zeus, stirreth the sea, and maketh the deep perilous.

Also in the spring may men sail; when first on the topmost spray the fig-tree leaves appear as the footprint of a crow for size, then is the sea navigable. This is the spring sailing, which I commend not, for it is not pleasing to my mind, snatched sailing that it is.

Hesiod, The poems and fragments, tr. A. W. Mair, Oxford, Clarendon Press, 1908.

[1] July–August.

It is interesting to ponder the gradual development from that stage to the holding of the high seas in any weather and finally as the best way of riding out a hurricane.

Indeed the organised study of weather began with the seaman. We may find glimpses of it in Dampier's book on the winds or in his *Voyages* already referred to, which cover a vivid description of a typhoon of the China Seas. They are closely followed by Halley's paper on the trade-winds and monsoons with a very celebrated map.

The middle of the nineteenth century supplied the careful investigation of the tropical revolving storms called hurricanes in the West Indies, typhoons in the China Seas and cyclones in the Indian Ocean north and south of the Equator, using a name coined for the purpose by H. Piddington to express the analogy of the air movement with the coil of a snake.

When the making of a weather-map was attained in the sixties of last century the analogy of the air movement led to the adoption of the name cyclone for the features of the maps which corresponded with the dips in the pressure records, easily recognised by every possessor of a barograph to-day.

On land the anticipation of impending weather is of economical importance for the control of transport. Exceptional flooding, exceptional warmth, fog and snow have to be provided against and all forms of outdoor sport are dependent upon weather, but it is not so easy to organise because the irregularity of the surface introduces difficulty which is only partially met by presenting a map for what is called sea level. It is often claimed, even at the present day, that the wise shepherd with his local knowledge and the fragment of the sky above him, is able to form a more accurate anticipation of coming weather than the forecaster in his office a hundred miles away with his conventional map; but the wise shepherd seldom expresses his views in writing.

Another class interested in weather is the soldier. In war time the need is very insistent. It used to be partially met by "going into winter quarters" to avoid the worst of the weather. Nevertheless, the weather has had a hand in the decisive battles of the world, both ancient and modern.

The introduction of gas warfare requires from the soldier a greater interest in the minor details of weather than that of anyone except the scientific agriculturist who has to bring the vicissitudes of weather into co-ordination with the physiology of the plant. And here it is to be remarked that the husbandman comes in almost last of all with his demands upon meteorology, when the resources of that science have to be reckoned in definite terms.

The subject has recently engaged the attention of the agricultural authorities who have arranged a special organisation to develop the study. Here, therefore, we may introduce some passages from an address to the agricultural section of a Conference of Empire Meteorologists in 1929.

AGRICULTURAL METEOROLOGY, A BRIEF HISTORICAL REVIEW

Weather has, of course, been a subject of fundamental importance for agriculturists from the beginning of time. It is, I suppose, of equal importance for all the beasts of the field, for the fowls of the air, the insects that infest it, the fish of the river, the lake and the shallow sea. The lower animals must have shared man's interest in the weather and in fact, if we are to believe famous Greek and Roman authors who wrote about the weather, lower animals are so skilled in knowledge of the subject that they notify coming changes which the lords of creation themselves are too slow-witted to divine; and many people are still of opinion that the very hips-and-haws can foretell seasons. But it is undeniable that the first knowledgeable application of meteorology was the planting of grain in a suitable place at a suitable time of year. That was the basis of the first co-operative movement, the dawn of our civilisation. So interwoven is the farmer's life with the knowledge of weather that one of the great difficulties in developing the application of meteorology to agriculture has been that, knowing as everyone does that the growth of plants and animals depends on sun and warmth and rain, we could not persuade the farmers to tell us how much sun, how much warmth, how much rain was necessary for a bumper crop. Agricultural meteorology is what the farmer knows and won't say.

But the farmer must have lived there for years, and his father before him, so that he finds a working knowledge of weather bred in his bones. At Washington once, on a visit to the Secretary for Agriculture of the United States, a Scotsman, which means a good deal in agriculture, I was much impressed by his telling me why it was necessary for the Ministry of Agriculture to have its Weather Bureau. At that time a great deal of the United States was virgin soil and the people who were to till it came from all parts of Europe and had not even that rudimentary knowledge of the relation of crops to sites and seasons that I have called the dawn of civilisation. If the range of climate and weather in the United States of America requires a Weather Bureau to prevent its citizens' efforts from being wasted, for want of knowledge of the weather and its meaning in beneficence and maleficence, what sort of Weather Bureau does the British Empire require in order to make the most of the soil that it possesses?

Our own Meteorological Office was started in 1854 for the sea and navigation, not for the land and agriculture. I should not be surprised if the Government of 1854 were of opinion that if anybody wanted to know anything about the weather and climate of the United Kingdom it would be better to ask a farmer than a Fellow of the Royal Society.

In 1877, after ten years of inquiry by a committee of the Royal Society into the relation between meteorological records and weather, a new Council entrusted by the Society with the control of the Weather Office began at once to take up the question of the application of meteorology to agriculture and public health. It was a very strong body of scientific men to whom the control of meteorological work was entrusted. Henry Smith, a versatile professor of mathematics at Oxford, who died too soon; Warren De la Rue, a pioneer in the study of the sun and a leading expert on the development of graphic representation; Francis Galton, who devoted his life to promoting the application of science to the service of mankind in common life; George Gabriel Stokes, the most distinguished of the successors of Newton at Cambridge, and Richard Strachey, Royal Engineer, Indian administrator, botanist and meteorologist, with Captain Evans, *ex officio*, as Hydrographer of the Navy, all famous men. For help with their agricultural scheme they consulted two other persons, whose names are well known, J. B. Lawes, the founder, and J. H. Gilbert, the first director of the Experimental Station at Rothamsted. We look to the same quarter for advice to-day.

In order best to serve their common purpose the council arranged to collect information about warmth and rain and sunshine, mainly contributed by voluntary observers, from representative stations in twelve districts of the British Isles and to publish every Wednesday the report of the weather of the week which ended on the previous Saturday, so that those who were interested might watch the progress of the crops in relation to the observed sequence of weather in the several localities. Subsequently small daily weather-charts for Europe and the Atlantic border were added, three for each day, two to show the march of pressure, the third to give the associated weather. Various modifications were introduced before the War brought in its changes. The first of the weekly numbers was issued at the beginning of February 1878—and the last at the end of January, 1928.

That was the expression of their ideas about practical climatology. Synoptic meteorology and forecasting belonged to a different department.

With the gradual accumulation of facts and figures the weekly report led first to the recognition of certain relations between the comparative success or failure of the cereal crops and the rainfall, warmth and sunshine of the previous seasons.[1] With the introduction of special statistical methods of inquiry for which Galton was largely responsible, in conjunction with corresponding investigations in other countries, particularly

[1] "Seasons in the British Isles from 1878," *Journal of the Royal Statistical Society*, vol. LXVIII, Part II, pp. 1–97, 1905.

in the United States, Russia, Sweden and Italy, agricultural meteorology has now a literature of its own. The topics which should be regarded as included in the subject may be represented by the following summary:

The sun and earth: energy, radiation, weather, life and growth. Soils and soil temperature, moisture and atmosphere. General physiology of plant life and growth. Shrubs, fruit trees and forest trees: climatic requisites for fruits and forests. Insect pests: influences of weather. Diseases of crop-plants: relation to weather-conditions. Meteorological instruments and records of weather. Climate and its biological significance. The phenological year: parallel columns of the sequence of weather and progress of crops. Weather forecasting and broadcasting: weather-maps. Statistical methods for predicting the yield of crops.

Synoptic meteorology and forecasting, often regarded as a sufficient expression of the farmer's interest in the weather, come in as scientific prediction of the future, parallel with the computation of the yield of the crops. And in fact they are not more closely related to agriculture than to any other outdoor occupation. Judging by what may be read in newspapers, cricketers are quite as sensitive as farmers to the special incidents of weather.

The efforts to trace the relation between the growth of crops and the conditions of the environment in which they grow are remarkable in that they have met with so little response from the rank and file of the students of agriculture. Special statistical investigations have been made; but there is very little regularity in the comparison of the progress of the crops with the sequence of weather. The comparison is really a part of phenology, and phenology, as expressed by a committee of the Royal Meteorological Society, deals with the occurrences which belong to wild nature, birds, insects, forest trees and wild flowers. Cultivated crops are not included.

And for this state of things meteorologists are to some extent responsible. In accordance with tradition their data are mostly arranged according to the months of the civil year beginning with 1 January, a very unsuitable cycle for the study of nature. The process of growth is nature's method of integrating all the influences of the environment upon the plant or the seed. Even for trees and shrubs the year's work begins with the buds which survive when the leaves have fallen in the chill of October. They have to be nursed through the winter for their opening in the spring. From that point of view a phenological cycle starting with the day after the bonfires of November 5th, in order to begin a quarter centred at the winter solstice, 21 December, followed by quarters centred at the spring and autumn equinoxes and the summer solstice, would be more suitable than beginning to count from January 1. Notes of nature's integration by growth must refer to successive periods, and it would be desirable to put in comparison with them, preferably week by week, the accumulation of sunshine, rainfall and warmth from the commencement of the phenological cycle.

The sequence of weather is itself a kind of growth, while the atmosphere which is also the environment of the plant is obeying the ordinary physical laws, and though we cannot imitate nature's integration we can note its progress both in crops and weather.

So little stress has been laid on this aspect of the subject that when it was decided in 1927 to issue the *Weekly Weather Report* as an annual volume March was taken as the commencement of a new year. The issue of 1931–2 is dated 8/32. An enterprising student of agriculture starting his inquiry with the beginning of his phenological year in November 1932, and ending with an audit of his success or failure in October 1933 would have to wait till August 1934, nearly a year after the passing of his agricultural audit, for meteorological information deemed officially to be necessary for his inquiry, and even then he would find the information guarded by the warning *Crown copyright reserved.*

When the meteorologists of the Empire come to grips with the subject of agricultural meteorology they will make a better show than that. Probably each Dominion will want its own phenological cycle, and why not? The recital of the list would tell us more about agriculture than is common knowledge now. A routine suggested for the home country comes with the Score of the Drama in Chap. III, and is dealt with in detail in Chap. VII.

AN ONLOOKER'S VOCABULARY OF GREEK WORDS, PREFIXES AND SUFFIXES (WITH SOME OTHERS) THAT STUDENTS OF WEATHER MAY WISH TO UNDERSTAND

It is perhaps the habit of using Greek words in scientific literature that causes the study of nature to be regarded in England as the speciality of a "highbrow". In Germany, where scientific knowledge is popular, the tendency is to substitute more homely words. There, a telescope has always been a "distance tube" and now a telephone is a "far speaker". The reader may take the opportunity of expressing the more difficult Greek in a more intelligible vernacular.

Prefixes or adjectives

Poly:	multiple, many
Baro:	weight of air per unit of area
Thermo:	warmth
Psychro:	chill
Hygro:	moisture
Hydro:	water
Hyeto:	rain
Atmo:	vapour or gas
Anemo:	wind
Helio:	sun
Bront:	thunder
Nepho:	cloud
Pheno:	what one can see in nature
Iso:	equality
Syn:	together
Geo:	about the earth
Hyper:	exceeding or above
Hypo:	less than or below
Hypso:	height or depth
Strato:	encamped, or (Latin) in layers
Tropo:	change by turning
Meteoro:	anything that is observed in earth or sea or sky
Chrono:	time
Stereo:	solidity

Suffixes

:meter	an instrument that measures
:graph	an instrument that writes or draws
:gram	a writing or picture
:scope	an instrument that seeks
:pleth	magnitude, quantity
:logy	discourse, talk or study
:bar, :therm, :hyet, :hel, :phene	as in prefixes
:neph	cloudiness
:chronous	on time
:optic	in sight
:static	at rest
:dynamic	under the influence of unbalanced force
:pycnic	density, closeness
:allobar	change of pressure
:steric	solidity or rarity
:meric	proportion or fraction
:akair	unseasonableness
:oid	like in form; aneroid is like a barometer but not wet

About Units

Meter or Metre:	a unit of length
Gram or Gramme:	a unit of mass
Dyne:	c.g.s. unit of force
Erg:	c.g.s. unit of work
Barye:	c.g.s. unit of pressure
Bar:	another unit equal to a megabarye
c.g.s.	centimetre-gramme-second

Deka:	ten times
Hecto:	a hundred times
Kilo:	a thousand times
Mega:	a million times
Deci:	a tenth
Centi:	a hundredth
Milli:	a thousandth
Micro:	a millionth

CHAPTER II

THE WATCHERS: WHAT THEY SEE AND WHAT THEY SAY

Methinks I see this hurly all afoot.

PANDULPH in *King John*

International symbols

SKY
UGLY THREATENING SKY

WIND
GALE
SQUALLS

PRECIPITATION
RAIN
PASSING SHOWERS
DRIZZLE
SNOW
SLEET
SOFT HAIL
HAIL
WATER SPOUT

ELECTRICAL PHENOMENA
THUNDER
LIGHTNING
THUNDERSTORM

ATMOSPHERIC OBSCURITY
FOG
WET FOG
MIST
HAZE
DUST STORM
ICE CRYSTALS
DRIFT SNOW (HIGH)
DRIFT SNOW (LOW)
VISIBILITY

GROUND PHENOMENA
DEW
HOAR FROST
RIME
HARD RIME
GLAZED FROST
SNOW LYING

OPTICAL PHENOMENA
SOLAR CORONA
SOLAR HALO
LUNAR CORONA
LUNAR HALO
RAINBOW
AURORA
ZODIACAL LIGHT
MIRAGE

International notation for cloud-forms

High cloud:	
Cirrus	Ci
Cirrostratus	Cs
Cirrocumulus	Cc
Medium cloud:	
Altostratus	As
Altocumulus	Ac
[Altocumulus-castellatus	Ac-cast]
Low cloud:	
Cumulus	Cu
Stratocumulus	Sc
Fractocumulus	Fc
Cumulonimbus	Cb
Stratus	St
Nimbostratus	Ns
Fractostratus	Fs
[Mammatocumulus	Ma-cu]
[Lenticular	Lent]

The symbol for thunder is also used to signify "thunderstorm in the neighbourhood of the station".

For the book of the play of wind and weather that should enable us to understand the drama we have to rely on our recollection of successive scenes in the pageant, on notes about the clouds in their courses and of the air-currents in which they are cradled and carried.

One important difference between Aristotle's meteorology and modern practice is that with the development of experimental science in the sixteenth century it became possible to supplement the visible information about the clouds by observations of the condition of the air as regards its warmth, and the sunshine that causes it, its drying-power—let us say its thirstiness—and another important but invisible feature of the atmosphere, namely pressure. Some sort of estimate of the force of the wind and the direction from which it comes would be a matter of common sense though it has taken centuries to refine it into a scientific record. The addition of a measurement of the amount of water that comes down as rain requires no great stretch of human ingenuity.

But naturally at first all these measurements were confined to that part of the air in which the observer could place his measuring instrument. It was only in the latter part of the nineteenth century that systematic observation of the upper air was organised. It is still inadequate; enough, however, to make it clear that the surface layer is full of tricks and pitfalls, the very last situation that a scientific investigator would choose if he wished to *understand* the drama, but the only one at which he is able to work at full power. It has been called the layer which nature uses in order to "camouflage" the operations which are natural in the upper regions of the atmosphere that we call the free air. As experience increases it seems that, in the future, no meteorological station will be regarded as complete unless it has some avenue leading to a knowledge of the layers above the treacherous surface.

The weathercock on the top of a tall spire pointing to the sky carries for meteorology a message of warning and of hope; the cock has always been regarded as a warning against misrepresentation and the upward pointing spire is an indication of where we must look for information that will give us the interpretation of the drama.

The observations which require no instruments and have been possible throughout the ages are indicated in the list of symbols at the head of this chapter. In order to furnish some details of the scene that presents itself to the observer it should be amplified by an illustrated list of the different forms of cloud, which is afforded by a cloud-atlas; and we must not omit the observer's own notes of the light and shade as well as the colour of the cloud-masses, a field in which no official guidance is provided but which has its own irresistible attraction. In these days the symbols and cloud-forms are international—that is to say they have the same meaning for the inhabitants of every country irrespective of language.

THE WINDS AND THE BEAUFORT SCALE OF WIND-FORCE

With regard to the winds, the mention of light airs, breezes, stiff breezes, gusts, squalls, gales, half gales, whole gales, storms and hurricanes would carry us back to the origins of the language of seamen; but only at the beginning of the nineteenth century was the language brought under the regulation of a scale by Admiral Sir Francis Beaufort; and only at the end of the same century were figures available to express the Beaufort numbers in terms of the real speed of the wind. In the days of sailing ships the wind-force was identified by the amount of sail a full-rigged ship could carry, and the limit—a hurricane—described as that which no canvas could withstand. Overleaf is a specification of the equivalents in velocity and pressure which every observer of weather may find useful, and of the incidents by which he can recognise the numbers. The equivalents in miles per hour are intended for the readings of anemometers with a very free exposure 30 ft or 40 ft above a level shore or open ground. Those for metres per second are related to continental exposures and every exposure really requires a special table.

The table gives a velocity of 75 miles per hour as the lower limit of the winds of "hurricanes" of the West Indies, or "cyclones" of the Indian Ocean, or "typhoons" of the China seas. Outside the tropics we may have extreme velocities in tornadoes, typified in the United States. Elsewhere records depend upon height, exposure and special circumstances. Our best example is a wind of 231 miles per hour at the Observatory of Mount Washington, New Hampshire (6284 ft.), on 12 April, 1934.

In the Antarctic, blizzards are recorded of more than hurricane velocity. The expedition of Sir Douglas Mawson to Adélie Land,[1] with winter quarters at Cape Denison, suggested an extension of the scale to meet the exigencies of the local blizzards to 100–200 miles per hour, and winds between 75 and 100 miles per hour were quite frequent in the transition months, May and October. Wind-forces 1 to 4 could be estimated by the ship's speed up to 6 knots, beyond that up to force 11 when she could carry only storm stay-sails, by the adjustment of canvas, and force 12, no canvas at all.

[1] J. Gordon Hayes, *The Conquest of the South Pole*, p. 147.

Admiral Beaufort's scale of numbers for the sailor's traditional notation of winds of different strength, with the probable equivalents of the numbers in terms of the velocity of the wind or the pressure exerted on a square foot of area facing it, also the adaptation of the same scale for use on land by identifying the effects of the several velocities or forces as recorded on an anemometer.

1	2	3	4	5
Sailor's Notation	Nos. 0–12	mi/hr	m/sec	Effects produced on land
Calm	0	0–1	0–½	Calm; smoke rises vertically
Light air	1	1–3	½–1½	Direction of wind shown by smoke drift, but not by wind vanes
Slight breeze	2	4–7	2–3	Wind felt on face; leaves rustle; ordinary vane moved by wind
Gentle breeze	3	8–12	3½–5	Leaves and small twigs in constant motion; wind extends light flag
Moderate breeze	4	13–18	5½–7	Raises dust and loose leaves; small branches are moved
Fresh breeze	5	19–24	7½–9½	Small trees in leaf begin to sway; crested wavelets on inland waters
Strong breeze	6	25–31	10–12	Large branches moving; whistling in telegraph wires; umbrellas difficult
Moderate gale	7	32–38	12½–15	Whole trees in motion; hard walking against wind
Gale	8	39–46	15½–18	Breaks twigs off trees; traffic disturbed
Strong gale	9	47–54	18½–21	Slight structural damage (chimney pots and slates)
Whole gale	10	55–63	21½–25	Seldom experienced inland; trees uprooted; structural damage
Storm	11	64–75	25½–29	Very rarely experienced; accompanied by widespread damage
Hurricane	12	Greater than 75	Greater than 29	— — — — — —

Beaufort number	0	1	2	3	4	5	6	7	8	9	10	11	12
Equivalent speed at 33 ft exposure mi/hr	0	2	5	10	15	21	27	35	42	50	59	68	Above 75
Equivalent pressure:													
(1) weight in pounds per square foot	0	0·011	0·08	0·28	0·67	1·31	2·3	3·6	5·4	7·7	10·5	14·0	Above 17
(2) millibars	0	0·005	0·04	0·13	0·32	0·62	1·1	1·7	2·6	3·7	5·0	6·7	Above 8·1

Col. 1 gives the sailor's traditional notation; Col. 2 the corresponding Beaufort number; Col. 3 the range of velocity indicated on a freely exposed anemometer in place of the description of the management of a ship's canvas; Col. 4 the corresponding scale in metres per second, adapted for continental exposures, rounded off to half metre steps; Col. 5 the effort to identify the Beaufort number by the wind's behaviour on land.

In the Meteorological Office a gale is understood to mean a wind of force 8 or more. Some years ago the Office endeavoured to substitute High Wind for Moderate gale in the scale, because the exclusion of force 7 from the category of gales by an adjective might be a serious embarrassment for an official witness in a court of law where decisions are taken as to whether damage from wind is or is not an "act of God", but the sailors' conservatism was too much for the effort. Moderate gale is restored to the winds less than gale force.

The normal relation between the Beaufort number, the velocity and the pressure of the wind with normal air density on an area facing it within the range of a square foot and 100 square feet, is expressed in an interesting way in the lower table of p. 82.

Notice, as a numerical curiosity, that the numbers for pounds per square foot at normal density are just double those of the millibar equivalent. A more formal apology for millibars as a measure of pressure comes later when we consider the pressure of the air that is not related to interference with its movement by an obstacle; but the duplicity of the pounds per square foot holds anyway and is worth remembering.

We might also find that there is a curious proportionality between the square of the wind's velocity and the cube of its Beaufort number which must somehow or other have its roots in the rig of old sailing ships.

During the War much use was made of a portable anemometer made of a swinging plate similar to that of Hooke's drawing on p. 123, and in these days, to make the force of the wind visible at aerodromes, a contrivance hitched to the top of a pole is used that swings out with wind and looks like a long stocking. Probably a long streamer of moderately heavy but flexible material would make a good indicator for ordinary meteorological purposes. In a way, a flag does that already to some extent. It might have a suitable code of figures based on its inclination, its oscillations and its ripples, but nothing of the kind seems to have transpired. Perhaps some day with the centenary of Admiral Beaufort in mind some Major-General or Air-Marshal may think it not unbecoming.

Orientation: the mariner's compass and the weathercock

We must say a word about what is called orientation, the identification of direction, and the mariner's compass. The cardinal points N, E, S, W are true directions; N means the direction of the pole star, S the position of the sun on the meridian, on the average for a whole year the position of the sun at noon, but at certain parts of the year according to the clock the sun is a little late in crossing the meridian, making up the difference by being a little early at other times. It is on that account that the evenings begin to get longer on 14 December a week before the shortest day, now St Thomas's day, and the mornings keep on getting shorter until the Epiphany on January 6; anyone who uses a sundial to tell the time will be familiar with the details. E marks the point at sea or on a flat plain where the sun rises on 21 March and 24 September, and W where it sets on these days.

These orientation-marks N, S, E, W will be found on the card of the mariner's compass, together with the intermediate points NNE, NE, ENE, etc., and the magnetic needle of the compass can swing to and fro over them. What we wish to emphasise here is that *the compass-needle does not point true north and south, the orientation by which the wind is recorded, but "magnetic" north and south.* At Greenwich in 1930 magnetic north, at which the magnet points, was 12½ degrees to the west of true north, ninety years ago it was 23¼ degrees. We might find it pointing 20 degrees west now, off the west coast of Ireland. It is different in different countries and changes a little with lapse of years. The angle between the magnetic north-and-south and true north-and-south is called the "magnetic declination" or "deviation of the compass", and has to be allowed for when the orientation of the compass is used to give direction. This warning is important in these days because the difference is often disregarded in setting up the orientation cross-marks on a weather-vane. In days gone by when the clock on the village church had to be set by sundial the incumbent took an interest in astronomy and the correct orientation was known; but now that the clock is set by information received by wire or wireless from Greenwich the village blacksmith or coppersmith sometimes uses a compass to set the wind-marks without thinking. A cautious meteorologist as he passes a weather-vane with orientation-marks, automatically takes out his pocket-compass to see if the points agree, and if they do he knows that the orientation is wrong. Almost within the precincts of the Science Museum in London is a church which is out of alignment with the north and south of the street. It is just possible that it has been so carefully oriented for Jerusalem with a compass that it actually points to Moscow.

THE WEATHER-GLASS: BAROMETER AND THERMOMETER

We pass on to consider the barometer for measuring the pressure of the atmosphere, the thermometer for measuring the temperature as an estimate of the warmth of air, and the hygrometer for measuring the humidity as an estimate of its moisture, or conversely of its thirstiness or drying-power. In quite early days after the instruments were invented in the seventeenth century they were combined to form what was known for many generations as a weather-glass. Later on we shall have to consider some of the dynamical and thermodynamical implications of the observations with these instruments, but at the moment we are referring only to the readings of instruments in common use with which readers will certainly be familiar.

In 1643 Evangelista Torricelli, who acted as amanuensis to Galileo in Florence in his blindness, discovered that the atmosphere would support a column of liquid mercury about 30 inches high, and no higher, in a tube closed at one end that had been filled with the liquid and inverted with its open end under the surface of mercury in a little cistern. The experiment was accounted for by allowing that the atmosphere exerted a "pressure" on the surface of the liquid, expressed in pounds weight per square inch about 14½ at sea-level, or some other unit of the same kind, and that the column of mercury in the tube *when there was no atmosphere on the top of it* exerted the same pressure on the liquid at the section of the tube at the same level as the surface in the cistern; and the pressure could always be computed from the difference of levels of the surface of the mercury in the tube and in the cistern. The height of the column was about 30 inches because with a liquid as dense as mercury, 13½ times as heavy as water, a column only 30 inches high was required. The experiment might have been made with water and often is in engineering practice; but the tube or its substitute has to be 13½ times the length of the column of mercury, say 34 feet.

We now recognise that the proper way of thinking about the atmosphere is that in the open air, or in any room which is connected with the open air even if only by a chink, every surface in the neighbourhood of sea-level carries a load of about a ton of air per square foot, or a kilogramme per square centimetre, with variations from time to time of about one-tenth of the whole load.

It thus becomes apparent that if one climbs up to heights above sea-level either on foot as on mountains or by balloon or aeroplane in what is called the free air, leaving part of the atmosphere below, the burden of pressure is correspondingly less and the top of the mercury-column in a barometer is lower. The effect is shown in fig. 40 p. 114.

The relation of the height of the column to the pressure of the atmosphere was illustrated in 1650 by Blaise Pascal who observed the height of the mercury-column on the top of the Tour Saint Jacques in Paris to be less than at ground-level, and afterwards much less on the Puy-de-Dôme, the cone of an extinct volcano near Clermont-Ferrand.

Not many years after it had been discovered that the height of the mercury indicated the pressure of the atmosphere on the surface of the liquid in the cistern, or more simply the pressure, the tube of mercury arranged as an instrument for measuring the pressure was called a barometer, beginning what is still a recognised practice of using Greek words when we talk about the weather—a just compliment to the Greeks and accepted internationally as being impartial to the moderns. The open end of the tube, bent upwards, was provided with a float counterpoised by a weight hanging over a wheel, and a pointer that ranged over a dial and showed the height of the barometer; next it became a weather-glass with the various kinds of weather set out on the dial at selected points which marked the heights of the column for "rain", "much rain", and "stormy" on one side, of "change", "fair",

The weather-glass. Front.

Fig. 32 *a*. Front, showing hygrometer at the top, thermometer above the small mirror, barometer scale with weather-legend, and at bottom a spirit-level.

The weather-glass. Back.

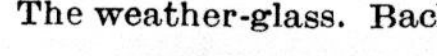

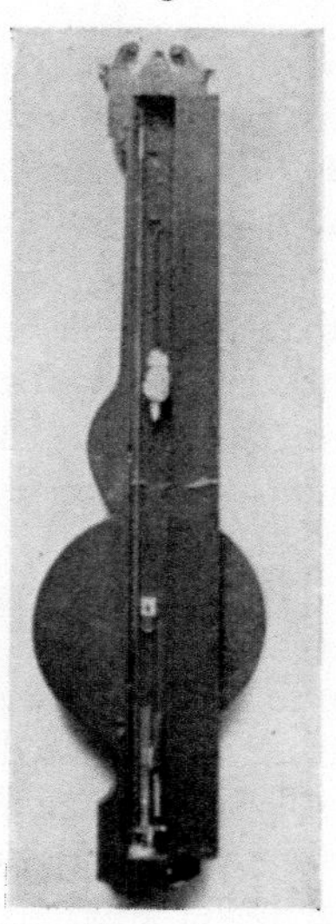

Fig. 32 *b*. Back, open, showing the long tube of the barometer and the short tube with float counterpoised over the central pinion.

"set fair", "very dry" on the other. A common form of the instrument is shown in fig. 32.

Almost at the same time the expansion of a liquid in a glass vessel was made available, first apparently by Galileo in his early days, to indicate the changes in warmth or coldness of the atmosphere. The vessel in which the liquid was enclosed was a glass bulb, spherical or cylindrical, with a long tube of very fine bore for its neck along which the surface of the liquid could move as its mass was warmed or cooled artificially, or by natural change in what is now called the temperature of the air. Mercury is again the best liquid for the purpose, but alcohol was an early favourite and is still used. A fixed mark had to be made to indicate the "zero position" of the surface of the liquid from which changes should be estimated; and the changes were indicated by a scale of tenths or twelfths of an inch, or some other, for the purpose of keeping account. Originally in the earliest Italian forms of the thermometer made in Florence the degrees were marked by beads of glass fused on the stem, each tenth one coloured.

In a volume of weather observations in 1710 with the columns headed *Barom.* (in inches and twelfths), *Therm.*, *Anem. Tempestas*, published in Frankfurt and Leipzig 1714, the reading was taken in twelfths of Paris inches up or down from a mark that might indicate the "mean" temperature. That process was soon developed into marking the freezing-point and boiling-point of water and dividing the space between the two into an agreed number of "degrees", 100 for the Centigrade thermometer, and 80 for that of Réaumur, the common scale on the Continent, each of which had the freezing-point of water marked zero, 180 for the Fahrenheit thermometer for which 32 marked the freezing-point and 212 the boiling-point.

Thermometers are glass instruments delightfully portable. It was easy to attach one to the case of the barometer-tube so that the weather-glass could show the temperature of the air in which it lived as well as its pressure. And for the accurate measurement of pressure a thermometer is necessary in order to make allowance for the variations with heat of the density of the mercury column of the barometer.

However, the thermometer does not count for much in the barometer as weather-glass, partly because the instrument, kept indoors, is not of much use for deciding whether it is freezing or not outside, whether it is the hottest day that ever was or not; but with the ordinary position of a barometer in the lobby of a house the changes of its temperature are only partly dependent on the changes in the weather and are seldom or never used by the householder for their ideal purpose of correcting the reading to that of standard mercury.

Hygrometer and dew-point

The weather-glass with its dial carries another addition (see fig. 32*a*), namely a humidity indicator—a hygrometer. A little pointer on a small dial passes from "wet" to "dry" or back again, according to the elongation or contraction of a hair attached to a spindle that carries the pointer to show whether the air in which the instrument rests is wet or dry, according to whether the water that it is carrying as vapour is near the limit of the possible or far from it. The limit of the possible is different at different temperatures; but the pointer tells the percentage for the time being any way.

The automatic movement of the beard of the wild oat was used as a pointer in the seventeenth century; hair was brought into the service by de Saussure at the end of the eighteenth century. Wet and dry, like the other descriptions on the weather-glass, may refer to the expectation of weather. In these days nobody pays much attention to it and that part of the instrument is apt to be out of order; but again it is a useful indicator of changes in the condition of the atmosphere and indispensable for testing the air of weaving sheds and factories—graduated from 0 to 100 it tells us the amount of water in the air as a percentage of the possible amount at the temperature of the neighbouring thermometer. For measuring humidity in these days a pair of thermometers is used, one called the wet bulb, which is covered with muslin and kept wet, and the other the dry bulb without any cover. The wet bulb is cooled by the evaporation of water from the muslin. The humidity is determined by a formula, too difficult for us, based upon the difference of the readings of the two. There are tables to help us out.

But the fact that humidity is itself important in weaving sheds may be accepted as an intimation that there is something peculiar about fibres that is worth attention and should find a place when we are thinking of the hygrometer and its use. The word itself is rather repellent, one of the most unfortunate results of using Greek names for the common things of everyday life.

It is a matter of common knowledge that water is carried about in the atmosphere in the form of an invisible gas. When it condenses into visible drops we see it as cloud and sometimes feel it as rain or snow or hail. The water to provide the visible precipitation is carried up as water-vapour—and, in days gone by, a vapour was distinguished from a gas from the fact that it might be condensed to a liquid by changes of pressure and temperature which were within the range of the laboratory practice of those days. But in these days the facilities for producing high pressure and low temperature have been so much extended that no gas escapes condensation by those who have the necessary apparatus, and the means of working it. So we cannot think of water-vapour as essentially different in character from the gases nitrogen and oxygen that make up at least 90 per cent of the bulk of the atmosphere in the lower levels. Only we happen to be living in an environment in which the transition of water to vapour on the one side and to ice on the other side and *vice versa* is an everyday occurrence in various parts of the earth and its atmosphere, whereas for the condensation of oxygen or nitrogen or the visible evaporation of mercury we have to realise conditions which are not within meteorological experience.

If when the air has acquired a certain amount of water as gas its temperature falls, water has to go out by condensation leaving behind only the amount which the air can carry; but, if on the other hand the temperature rises, then the air becomes capable of carrying more water than it has, and is called "dry". Dry has so many different meanings and so many implications that thirsty is a better description of the air which can take up more water, and will take more, if it is within reach. Thirsty it will remain until its temperature falls or it finds some water in a liquid form to satisfy its thirst.

Here comes in the peculiarity of vegetable and animal fibres, cotton or hemp or linen or wood or hair. Exposed to the air they take up moisture at temperatures below saturation and alter their shape and volume by doing so. And curiously enough the condition of the fibre depends not on the amount of water contained in the air which forms its environment but on the fraction by which the actual quantity is related to the quantity at saturation at that temperature.

This curious property depends upon the capacity of the fibres to condense water under specified conditions in the pores, and the force exerted in doing so is remarkable, as every yachtsman knows, for a rope which is made taut when dry will tear something down if it is moistened, and cordage is as sensible to the dampness of the air in the cold Antarctic as in the torrid tropic. The amount of water involved is very different but the fibre is always satisfied with its fractional share and not otherwise. So hair or any vegetable fibre can make a sensitive hygrometer. Paper is used in a very recent and very simple form of dial-hygrometer.

It is very remarkable that so long as it retains its structure every substance which has formed part of a living organism—wool, hair, leather, whalebone, cotton, hempen rope—is sensitive and similarly sensitive of the humidity of the air and acts automatically as a hygrometer. With some provision for dissipating the moisture absorbed they may act as drying agents. In arranging for a special room at the Cavendish Laboratory for electrical experiments, Clerk Maxwell provided a roller-blanket, like a roller-towel, to be driven by clockwork over a drum heated by gas, in order to keep the room continually dry during experiments, because the electricity to be investigated has an awkward way of leaking along any damp surfaces. His successor, Lord Rayleigh, used blankets dried from time to time to carry away the moisture of basement air.

Other creatures too have some cleverness in respect of moisture. In the *Proceedings of the Royal Society* there is a romantic story of locusts from a biologist in Palestine. He was curious to know how the insects, who had to suffer a daily loss of water by evaporation in order to live, managed to supply themselves with the necessary water in the desert where no water was. The explanation proved to be that dried herbage was blowing about the desert hopelessly dry in the hot hours of the day when the water-vapour of the air was a very small percentage of a very high possible, but in the early morning hours when the air is cooled it is very near its dew-point; so then the herbage could take up the moisture from the damp air and the locusts chose that time to eat it and appropriate the water. We may just wonder whether Maxwell or Rayleigh would have thought of that if they were themselves thirsty. In any case what we have set out explains the domestic virtue of airing woollen clothes or blankets by driving out the water which the fibres acquire in an atmosphere which is nearly saturated, that is, not far from its dew-point.

A word or two about this dew-point. We have said that if the temperature of the air goes below its saturation point the water will collect into drops; but it is also true that if the air is in contact with a surface sufficiently cold the water will form drops on the surface. That is the process of the formation of dew, and it provides another historic method of specifying the humidity of the air by artificially cooling a polished surface, silver or glass, and noting the temperature at which dew begins to form on it. The cooling can be accomplished by a stream of ice-cooled water (in the hygrometer devised by Mr G. Dines) or by the evaporation of ether (in the Daniell or Regnault hygrometer).

By this process deposits of water are formed also on cold walls and on the outside of glasses which contain liquid cooled by ice. Some readers may remember the question of the Indian nabob who found an abundance of water deposited on the outside of his glass and asked, "Why does the water come through the glass and not the brandy?" The answer implies a considerable insight into the physical behaviour of the atmosphere. The dew-point is a point of fundamental importance.

THE USE OF THE WEATHER-GLASS FOR FORETELLING WEATHER

With the addition of a level to show when the case is properly vertical the barometer as weather-glass is now complete. In modern times a barograph which makes its own record of pressure is in common use instead of or in addition to a mercury barometer.

We may add a little talk about it after the fashion of the Guides to Science in vogue at the time when the barometer was at the height of its reputation as a weather-glass.

Q. Why is the reading of the barometer the best indicator of the prospect of changes in the weather?

A. Because change in the barometer is closely associated with changes in the direction and force of the wind.

Q. What sort of changes are indicated by the barometer?

A. If the barometer is falling with a south-westerly wind, rain is quite probable and the wind will change in the course of the day to west or north-west as the barometer rises, etc.

Q. Why are the readings of thermometer and hygrometer on the weather-glass generally disregarded?

A. The readings of both these instruments are more largely dependent upon the time of day, than upon changes in the weather, as in ordinary circumstances the air, indoors or out, goes cold at night and is warmer in the early afternoon, and generally speaking when the air is warm it is dry and when it is cold it is moist. So the use of these parts of the weather-glass requires more personal judgment than the barometer.

Q. Is the reading of the barometer a sign of weather to come?

A. No.

Q. Why?

A. One of the most effective answers to this question turns on a question asked of the Meteorological Office by a naval officer during the War as to why during a week when there had been quite important changes in wind and weather on his ship his barograph had kept a steady line throughout. The answer is that weather-changes depend not on a single barometer but on the relative variation of pressure in different parts of the whole region. So if the barometer changes something will probably happen, but if it doesn't change things may still happen. An illustration is given in *The Weather of the British Coasts*.

Q. Are the legends inscribed on the weather-glass, "change", "rain", etc. no good?

A. It is not so much a matter of good or no good as of sufficient or insufficient. No one need be surprised at stormy weather when the barometer is down to 28 inches or set fair at 30·5; but the converse is not always true. When the weather is stormy the barometer is not always at 28 and so with the others. Moreover, the reading of a barometer depends upon the height of its station. It loses a hundredth of an inch for every ten feet of height, a millibar for thirty feet. That has to be allowed for when barometer readings from different stations are compared. Weather-maps show pressure *reduced to sea level* by a conventional formula. The picture of the land barometer in fig. 35 shows how the outer dial can be adjusted to allow for the reduction; but the practice may easily be pursued so far that the local pressure loses any meaning it might have. Admiral FitzRoy, writing for seamen, substituted rules about the meaning of changes in the pressure instead of the readings of the single position. Some barometers with a visible column of mercury carry FitzRoy's rules as legends. With the introduction of the weather-map they have lost their hold.

Q. How could an observer make better use of his barometer?

A. By being in touch by telephone with a number of friends 50 or 100 miles away and ringing one another up day by day to say what their barometers are doing. That is in effect the construction of a weather-map, and in these days Government provides an efficient means of making one.

Q. Can anybody make a weather-map?

A. Yes, anyone who has a wireless set and can take down messages in Morse code.

Q. Could he foretell the weather when he had made the map?

A. We shall see about that later, in the meantime he might try to explain the weather of yesterday.

SPECIAL UNITS OF MEASUREMENT

Millibars

There are many forms of barometer. We have based our description on the mercury-barometer in which the reading is determined by the difference of level of the mercury in the top of the tube and in the cistern, where the mercury is exposed to the pressure of the air; the opening by which the outside air communicates with the cistern may be the merest chink, a small hole covered with a pad of cloth will do. The tube may be straight or bent, it is only the difference of level that counts; but if the difference is to be measured *along the tube* the tube must be vertical. Vertical means the direction of the plumb-line.

The difference of level is expressed sometimes in inches, sometimes in millimetres. In English-speaking countries the inch has been in common use, except in the laboratories of the scientific where the millimetre is used as it is in the rest of the world. But it must be remembered that the reading to be recorded is practically never the figure that is *read* on the barometer. That has to be corrected for the temperature of the mercury, the temperature of the scale and also for latitude. In the final result the inches have to be true standard inches and the millimetres true standard millimetres as defined for us by the Board of Trade and in practice the inches or millimetres marked on the scale are never "standard" when they are looked at.

That fact makes it very remarkable that there is any hesitation about carrying the correction further and expressing in millibars the pressure indicated on any mercury-barometer. A millibar is a unit of pressure belonging to a system that is called absolute because the selection is dictated not by arbitrary choice but by its relation to other units. The name millibar implies that it is the thousandth part of a bar, and the bar in this case is a million times the unit of pressure on the c, g, s (centimetre-gramme-second) system, selected originally in order to systematise electrical measurements. After some uncertainty in scientific circles the meteorological practice has been confirmed for international use. So for us it is sufficient that we should remember that a thousand millibars or a million baryes correspond very nearly with 29½ inches, 750 millimetres of mercury at standard temperature, and also very approximately with 1019 centimetres of water.

The millibar is in fact very nicely represented by the pressure of 1 centimetre thickness of water. In middle latitudes it is within 2 per cent of accuracy, so we should get a good working idea of pressure in millibars if we used a water-barometer and measured the height of the column in centimetres. And that really gives a much better idea of the meaning of pressure than anything that is got by quoting inches or millimetres. The pressure of the atmosphere about sea-level is roughly a ton per square foot. It may vary from time to time by having a hundredweight, or even two, put on or taken away.

A ton per square foot is not far different from a kilogramme per square centimetre, and a gramme per square centimetre with some allowance for variation of latitude is 0·981 millibar. So if we carry the name of millibar through our expression of pressure we are much closer up to the notion of the force on a square foot by remembering that a cubic centimetre of water weighs a gramme than we are by remembering that a cubic foot of water weighs 62·3 pounds.

Let us sum up the situation for those who have little time to examine details. When you see a column of water in a tube think of a millibar for each centimetre, approximately the thousandth part of the normal atmospheric pressure at 106 metres above sea-level. When you see a column of mercury think of a centimetre of it as one seventy-fifth part of the same atmosphere or an inch as ten two-hundred-and-ninety-fifth parts of the same.

The important point is not whether you *read* a mercury-barometer in inches or millimetres or in something else but that you should reach millibars in the end. For scientific purposes the end has to be reached by corrections by means of tables made for the purpose, and the final result can be given in millibars just as easily as in inches or millimetres. Fig. 33 shows the scale of a millibar-barometer and the note at the top specifies the condition when the reading needs no correction; in any other condition a correction is needed.

Millibars are now employed in the telegrams of exchange of daily observations and in other international enterprises.

So far, we have been regarding the mercury-barometer as the measuring instrument. Other forms of barometer are called aneroids—they

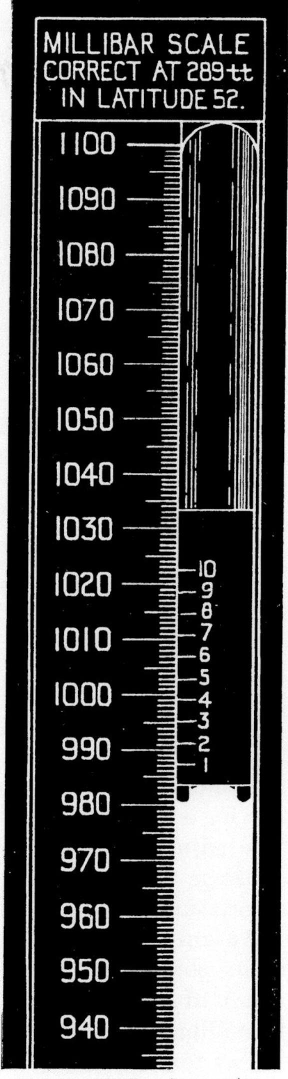

Fig. 33. Millibar barometer.

are mechanical contrivances and until experience verifies their behaviour they ought to be suspected of a propensity for changing their readings through imperfection of the delicate mechanism. The mechanism consists of a hollow metallic box or capsule nearly exhausted of air with a lid fluted to make it flexible and a spring inside to keep the lid in a position which may be altered by altering the pressure on the lid. A lever is mounted with a needle to show on a dial the variation in the pressure on the lid. The position of the needle depends on the balance between the force of the pressure on the exterior and of the spring in the interior of the capsule. Variation of gravity does not affect it, so no correction is required for latitude.

Fig. 34. Sea barometer.

Temperature does affect it and can be corrected automatically by leaving a suitable amount of air in the capsule when it is being exhausted. The adjustment is very delicate. It can be tested by comparing the readings with those of a mercury-barometer under the same variations of pressure.

An aneroid is very convenient in many ways; it can have the scale of its pointer graduated to show the reading of pressure in millibars, standard inches and standard millimetres simultaneously. The transition to millibars is merely one of scale.

Two forms of aneroid are shown—one for use at sea in fig. 34 gives the reading in millibars with inches on an inner circle. It also suggests the correction, never very large, for the height of the instrument above the level of the sea.

The other (fig. 35) for use on land shows an outer ring with a scale numbered in centibars 93 to 105 (a centibar is 10 millibars) that can be set for the height of the station above sea-level, a pointer and little scale of feet show the proper adjustment. And in the compartments between the numbers are shown (1) the normals for Shetland and for Jersey and (2) the number of occasions in the year when the pointer has been in each compartment or in the more extreme compartments the number of years one may have to wait for a record.

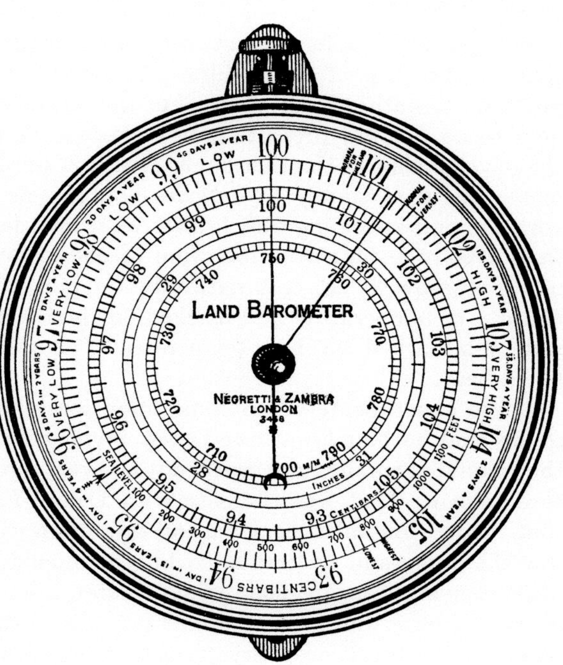

Fig. 35. Land-barometer.

Temperature, its measures and degrees. Absolute limit of cold

As with pressure there are many forms of barometer and of scales, so it is with temperature—there are many forms of thermometer based on changes produced by heat, generally in a liquid, mercury or alcohol, but sometimes in solids as the bimetallic recorders, or in gases as the hydrogen thermometers of the physical laboratory; and many scales,

Fahrenheit, Centigrade or Celsius, Réaumur, absolute Fahrenheit or absolute Centigrade or thermodynamic, tercentesimal, Kelvin kilograd.

When the barometer was invented the experiments disclosed a vacant space at the top of the column of mercury where there wasn't any pressure, so naturally the "zero" of the barometer-scale is the point of *no pressure*, and any pressure to be measured is reckoned from that. With the development of the thermometer things were different. The thermometer was used simply to indicate how hot or how cold a body was, with the understanding that heat flows from a hot body to a cold body, from a body of higher temperature to one of lower. The freezing-point of water, now 32° F, was naturally thought of as one of the salient features of temperature and marked on the thermometer. Other points were also marked, 55° F as temperate, 76° as summer heat, and 98·4° as blood heat.

If the experimenters had come upon something with a temperature so low that nothing lower could exist, no more heat could be got from it, they would naturally have marked it as a real zero and measure temperature from it. That was the case with Daniel Fahrenheit who thought that the freezing-mixture of ice and salt, often used still for making ices, was the limit; and he graduated his thermometer from that point. He could then mark the freezing-point of water, the temperature of a mixture of ice and pure water, as one fixed point, and the temperature of water boiling under the atmospheric pressure as another. He divided the range between those two points into 180 degrees, corresponding with the ordinary division for angles, and his freezing-point was found to be 32 degrees above his zero. It was not until the middle of the nineteenth century that it was made clear that there is an absolute limit, which could be calculated theoretically, where there is no more heat to be transferred and below which therefore temperature cannot go. It is approximately 459 degrees below the Fahrenheit zero and 491 below the freezing-point.

But in the meantime the importance of the freezing-point of water, implying all the economic and climatic difference between frost and thaw, was chosen as a zero for the Centigrade scale, introduced by the

Swedish chemist Celsius, with 100 for the boiling-point; and also by Réaumur, a French philosopher, who marked the boiling-point 80 instead of 100. Anything above the boiling-point came into the scales without difficulty and anything below the freezing-point could be measured in degrees corresponding with equal lengths on the thermometer but called negative and marked algebraically with a minus sign. On the Centigrade scale the theoretical absolute limit works out at within a tenth of a degree of -273 degrees. For those who move in mathematical circles there is much virtue in a minus sign, but it is apt to bewilder the ordinary person who tries to think for himself.

Persistent experiment on the production of low temperatures has led to remarkable results. The gas helium, isolated at the end of last century, has been liquefied at a temperature of -269, only 4 degrees from the absolute limit, and a temperature has been measured only one quarter of a degree from the limit; and within the range between that and the temperature which Fahrenheit regarded as the limit all the gases of the atmosphere have been liquefied and solidified and liquid air has become a commercial product.

On the other hand there is no limit yet ascertained for a maximum temperature. Measurements have been made up to about 3400 Centigrade; the temperatures of the filaments of electric lamps are in the region of 2500. The sun is estimated to have a surface temperature of 6000, and modern astronomers do not hesitate to write about internal temperatures of millions of degrees in the stars.

The temperatures of the air which occur in meteorological practice all lie between the limit of 93 Centigrade degrees below the freezing-point ($-135°$ F), a temperature observed in the upper air of Java, and 57 degrees (135° F), a temperature observed at the surface in Greenland Ranch, California, a range of 150 degrees (270° F) within the wider range of three or four thousand degrees of the experience of applied physics. Higher temperatures could be obtained by judicious exposure of thermometers to the sun, as high presumably as the boiling-point of water, but they are not counted in the normal meteorological forms.

This remarkable restriction of meteorological temperatures within a narrow range, with its meaning for the existence of life on earth, is worth

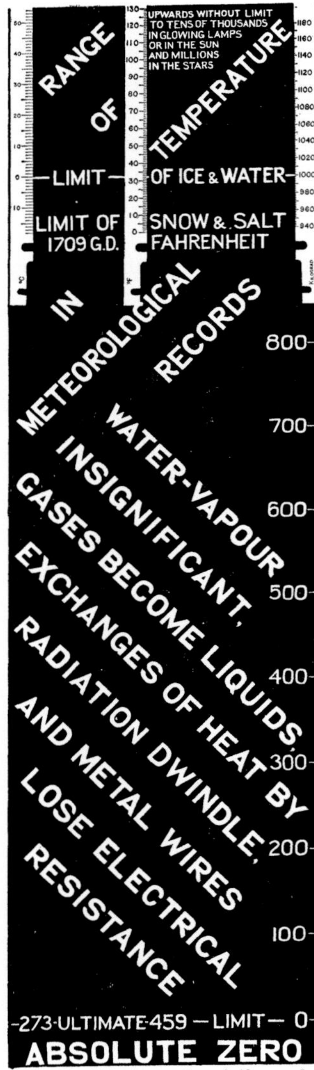

Fig. 36. Photographs of the scales (McAdie) of a meteorological thermometer with the extension downwards to the ultimate limit.

attention. We have made a diagram (fig. 36) to exhibit the limitation with some salient features of temperature for comparison.

With the absolute limit so clearly defined it is reasonable to regard it as a real zero for the measurement of temperature and to discard the use of minus temperatures which must confuse the inexpert reader. Various efforts to achieve this advantage in meteorological practice have been tried. Temperature measured from the absolute limit in degrees of the Fahrenheit scale have been called absolute temperature Fahrenheit, and with the Centigrade degrees absolute temperature Centigrade, which is the Kelvin absolute scale, sometimes indicated by K, or "absolute temperature" or simply A.

We have explained that the absolute limit though very nearly approached has never yet been reached, the final degree is a matter of refined calculation; but meanwhile the reading in the Centigrade scale with 273 added is often called absolute temperature, and for all meteorological purposes may be so regarded; but if we use the adjective absolute we are always liable to challenge from the thermodynamical expert and more than a single lecture would be required to explain how absolute temperature is lawfully obtained from a reading of the thermometer that one buys in a shop. So in self-defence we call the reading of the ordinary thermometer, amplified by the addition of 273 to the Centigrade scale, the tercentesimal temperature with tt as its symbol, and no degree mark, thereby placing ourselves on the side of the scientific angels who think dynamically, without pursuing all their intricate reasoning. For them at any rate a *negative* temperature is quite out of the question, a thing imagination boggles at.

One step further has been proposed by Professor McAdie. We are all unwilling to sacrifice the clear indication of the freezing-point that arises from using it as zero. With the Fahrenheit scale we are often in difficulties in the newspapers as to whether a temperature of 10° below zero is intended to mean a temperature of −10° F or 10 degrees of frost. Professor McAdie suggests that using the absolute limit as zero we might call the freezing-point a thousand. Not much practical use has been made of the suggestion but it may fairly be looked upon as the scale of the future. Or perhaps a freezing-point of 500 might be better.

Our diagram shows the range of meteorological temperatures from the 700 mark somewhere between the absolute limit and the freezing-point to 1200 at the top of the diagram, 200 steps above the freezing-point. This scale is called the Kelvin kilograd scale and its symbol is kd; but it might with advantage be kk. Prosit!

THE CONTROLLING INFLUENCE OF TEMPERATURE OVER HUMIDITY AND RADIATION

The physicists have introduced temperature to us as that which tells us how hot or how cold a body of air or water is and thereby indicates in which direction heat will flow when bodies of different temperatures are in contact. That is of course a very important consideration; but, for the study of weather, temperature means much more than that on account of the extraordinary properties of water, the substance on which all life depends. At temperatures which are of common occurrence on this particular planet the material can exist either as solid ice or liquid water and all the time the surrounding air will be carrying some load of water in the form of a gas or, to use a name which was in vogue before the word gas was invented, as water-vapour.

It is temperature which decides whether the form shall be ice or water. At what we call the freezing-point the two exist comfortably together, but for a temperature above that point the ice becomes water, below it water becomes ice. And in any case the air above contains a certain amount of water as vapour or gas.

We have represented the pressure of the atmosphere as the load of air contained within a vertical column of specified area. At sea-level the ordinary load is about a ton per square foot or a kilogram per square centimetre, and the barometer which measures pressure asks no questions about whether the air of the load is wet or dry, only how much the load is. Lighter air would have to be continued upwards to a greater height in order to give the same load—that is all. Temperature is more particular. Generally speaking air consists almost entirely of a mixture of the gases nitrogen and oxygen. There are however small amounts of argon, helium, neon, krypton, gases which were unknown to science fifty years ago. These are all called permanent gases because they only turn to liquids when the temperature gets near to absolute zero. Something between four parts in a thousand and two parts in a hundred may be water-vapour and particles of soot or dust or even water are visible as motes in the sunbeam. So generally speaking the permanent gases carry the load that is indicated by the barometer; but if there is water at the base of the column and things are left in peace the water insists upon taking a definite share of the load defined by the temperature, and for that purpose turns part of itself into gas or vapour until its share of the load is adjusted to what temperature dictates.

In the free air it is not easy to find examples of the complete assertion of the rights of temperature, but in the protection of a laboratory where the air and water can be left together in peace in a closed vessel the gradual evaporation of the water and its diffusion over the closed space can be watched and the share of the load which is dictated by a particular temperature tabulated. At the boiling-point of water the whole barometric load is carried by the vapour. The first recognition of this assertion of the rights of temperature over water-vapour was one of the conspicuous achievements of John Dalton, a schoolmaster and chemist of the early nineteenth century.

In the closed space the share of the load which the water insists on carrying is just added to the original measure of the load of dry air, but in the open air where there is freedom of movement the evaporated water is allowed to displace the equivalent in dry air without disturbing the total load measured by the barometer.

Air in which the amount of water-vapour has reached the limit indicated by its temperature is said to be saturated with water and experiment has provided us with the measures of the share of the load which is borne by the water-vapour in saturated air at different temperatures within the meteorological range and beyond that up to the temperatures of steam-boilers and other strong vessels.

We can express the limit of the amount of water-vapour which any space, a cubic foot or cubic yard or cubic metre, can carry when it is saturated, by the amount of the vapour, or by the part of the load of pressure which it carries. Water-vapour is lighter than air with the same volume, temperature and pressure, so in the atmosphere saturated air is lighter than unsaturated air, moist air is lighter than dry air under the same conditions of temperature and pressure.

The state of the air in respect of water-vapour can always be expressed by the portion of the pressure which the water-vapour bears, that is the vapour-pressure at the time, or by the amount of water in grains per cubic foot or grammes per cubic metre. These can be set out in a table if we can determine the temperature to which the air must be reduced without altering its barometric pressure in order to make it saturated. Then we can express the load carried by the water-vapour as a percentage of what was possible at the temperature and so characterise the air. This is the common way of describing the state of the air in respect of water-vapour; the percentage gives the relative humidity to which we have already referred.

Just as with temperature there are many different ways of expressing the condition in numbers, so with humidity there are many ways of expressing the relation to the control of temperature.

These facts about water-vapour and its limits are set out in the table of p. 104. They are among the most important of the facts pertaining to the study of weather because they provide the key to the formation of cloud, rain, dew, hoar-frost and in fact nearly the whole of weather.

The way of it is like this. We have allowed that temperature rules the quantity of water-vapour that air can contain and the quantity

TEMPERATURE'S CONTROL OF THE ENERGY OF RADIATION AND OF THE POSSIBLE AMOUNT OF WATER-VAPOUR IN A LIMITED SPACE OF SATURATED AIR

Expressed in systematic units based on the metre or centimetre, gramme and second

Temperature in several units			Radiation from black surface	Water-vapour		
				Pressure	Density	Load in 1 kg of dry air,
°F	°C	tt	watts/m²	mb	g/m³	g
90	32·2	305·2	496	48·2	34·1	30·8
85	29·4	302·4	479	41·1	29·5	26·0
80	26·7	299·7	461	35·0	25·4	21·9
75	23·9	296·9	444	29·7	21·7	18·7
70	21·1	294·1	426	25·1	18·5	15·7
65	18·3	291·3	411	21·1	15·8	13·1
60	15·6	288·6	396	17·7	13·3	10·9
55	12·8	285·8	381	14·8	11·2	9·1
50	10·0	283·0	366	12·3	9·4	7·6
45	7·2	280·2	352	10·2	7·8	6·3
40	4·4	277·4	339	8·4	6·6	5·2
35	1·7	274·7	326	6·9	5·5	4·3
32	0·0	273·0	318	6·1	4·8	3·8
30	− 1·1	271·9	313	5·6	4·3	3·4
25	− 3·9	269·1	300	4·4	3·7	2·7
20	− 6·7	266·3	287	3·5	2·7	2·2
15	− 9·4	263·6	275	2·8	2·3	1·8
10	−12·2	260·8	264	2·2	1·8	1·4
5	−15·0	258·0	253	1·7	1·4	1·0
0	−17·8	255·2	242	1·3	1·1	0·8

For instance: On 5 June 1933 the thermometer in the author's study read 70° F, 21·1° C, 294·1 tt and my "paper-hygrometer" said that the humidity was 71, meaning that the amount of water in the air was 71 per cent of the amount corresponding with saturation at 70° F.

So I know from the table that the air of the room, 140 cubic metres (5000 cu. ft) weighing about 170 kg (one sixth of a ton) contained about 1·8 kg (4 lb) of water-vapour. The amount possible was 2·6 kg (6 lb). The air in "our street" between mine-and-my-neighbour's houses and the pair of houses opposite, about 20 metres, 66 ft, each way, along, across and up, was carrying about 100 kilogrammes, one-tenth of a ton, of water. The pressure borne by the vapour was 17½ mb (·52 in or 13 mm of mercury) and might have been 25 mb (·75 in or 19 mm of mercury).

Vapour would have begun to condense if the air had been cooled to 60° F, 15·6° C, 288·6 tt, that being the *dew-point*, the temperature of saturation with the vapour-pressure quoted.

Also the Radiation column tells us that with a perfectly clear sky the 400 square metres (480 square yards) of road would have been losing heat by radiation to the sky at something less than 170 kilowatts, 170 B.T.U. per hour, worth in the neighbourhood as electricity at 1*d.*, 14*s.* 2*d.*, or as coal gas, 5·8 therms, at 8·6 *d.* per therm, 4*s.* 2*d.*

which is carried by any specimen of air depends on its past history. And the air which carries it is liable to changes of temperature. If for any reason it goes downwards through its environment to higher pressure it becomes warmer automatically and it also becomes warmer if its component parts receive heat by radiation from earth or sun. On the other hand if for any reason it goes upward through its environment it becomes cooler automatically and it also becomes cooler if its component parts lose heat by radiation to its surroundings.

We need not pay much attention to the warming, that only makes the air drier; but the cooling makes history of a more definite kind. A certain amount of cooling brings the carrying air to a state of saturation and further cooling takes the air beyond saturation and causes the superfluous water-vapour to be deposited as a cloud of drops of water or particles of ice in the free air if there are nuclei on which drops can form as there generally are, or as dew or hoar-frost if the ground or some other solid object is the agent by which the air is cooled.

We have mentioned radiation and must refer to it in order to point out that that also is controlled by temperature. One of the mysteries of nature is that every body in the universe has the faculty of transmitting energy from itself to every other body within sight by waves which travel across the intervening space at the rate of 186,000 miles a second if the space is empty, rather slower if it is occupied by air, water or glass which is still transparent.

The phenomena of radiation are extraordinarily complicated; they include everything that can be seen and all the transmission of warmth that cannot be seen. The one law that is of fundamental importance in the study of weather is that anything in a transparent environment that looks black at ordinary temperatures is really radiating heat across its environment and becoming warmer or colder according to the balance of account between it and the things it can see, and every one of them is radiating according to its temperature. Doubling the temperature as measured from absolute zero means sending out sixteen times as much energy by radiation.

This also is tabulated in the column of the table which is marked with the heading Radiation.

In practice hitherto, different units have been employed in different countries for the expression of temperature of air, for pressure, density of air or vapour and for the load or mass of water-vapour that can be contained within a given space. There would be great advantage in the adoption on an international basis of a system of units that would be common to all countries. The physicists who have to deal with electricity and magnetism have already for many years adopted units for those sciences based upon the centimetre, gramme, second as units of length, mass and time. Units based on the same system are also available for meteorological science; the adoption of those units is urged for obvious reasons. In setting out the table of the influence of temperature we have used that system for all quantities except temperature about which international agreement has yet to be reached. We have accordingly quoted temperatures in the Fahrenheit and the Centigrade scales and also in the tercentesimal scale running from 273 Centigrade degrees below the freezing point of water, a point sufficiently near the absolute zero of temperature for the tercentesimal temperature to be regarded practically as absolute.

* * * *

As we are now proceeding to consider the details of practice of a meteorological station or observatory we may set out here with advantage the alphabetical notation for weather removed from p. 48 of the original edition.

Alphabetical notation for weather (see p. 30)

Blue sky	b	Ice-storm	i	Snow	s
Cloudy sky	c	Rime	j	Thunder	t
Drizzle	d	Lightning	l	Ugly sky (threatening)	u
Wet air	e	Mist	m	Unusual visibility	v
Fog (visibility less than 1100 yards)	f	Overcast sky	o	Dew	w
		Passing showers	p	Hoar-frost	x
Gloom	g	Squall	q	Dry air	y
Hail	h	Rain	r	Dust-haze	z

With the development of modern methods of transport on land and sea and in the air, visibility has become so important that a whole scale has been assigned to it with a code to specify all the stages between f and v.

A METEOROLOGICAL STATION

In ordinary meteorological practice we do not combine pressure, temperature and humidity in a weather-glass; we keep the instruments separate; the barometer indoors with as constant a temperature as may be, two thermometers in a louvred box, called a screen, in the open air, so that they may be properly exposed to the air but protected from the sun and rain; the two together, wet and dry, provide a means for measuring the humidity of the air. That is regarded as more generally trustworthy than the variation in a hair which may get dirty without our noticing it. The wet bulb has been known on occasions to suffer from the same complaint.

Besides the wet and dry bulbs the screen contains a thermometer that is arranged to leave a mark at the highest point that the mercury has reached since it was set, and another one to leave a mark at the extreme of its excursion downward. Maximum and minimum thermometers these are called. They are set at least once a day, the minimum in the morning and the maximum in the evening.

With these instruments is always associated a rain-gauge to measure the amount of water that falls as rain upon a specified area. For catchment area British instruments have the opening of a can either 8 inches or 5 inches in diameter, which leads to a funnel that conducts the fallen water to a bottle or smaller can. Once a day at a fixed hour the observer comes and pours the collected water out of the collector into a measuring glass that tells him what the depth of water would have been on the catchment area.

There are certain details about exposure, which has to be good but not too good. It must be in an open situation free from the screening of trees or buildings; but it may want protection against the wind, because if the wind blows very strongly it may blow the raindrops across the catchment without letting them fall in. An *Observer's Handbook* tells us all about that.

The rain-gauge is by far the most popular instrument in all the meteorological equipment for systematic observation. There are some five thousand regular observers in the British Isles alone with the majority of whom measuring rainfall is a voluntary service to the community. The results are of vital importance in the consideration of water-supply and drainage; questions of increasing urgency with the increasing concentration of population in towns.

There may be actually more barometers and thermometers sold than rain-gauges; few houses of even moderate pretensions are without them; but for scientific purposes a thermometer needs a special kind of exposure and very few people trouble to get a reading of the barometer that would pass muster as representing what a meteorologist understands by the barometric pressure. Still, many stations that contribute observations regularly to the Meteorological Office are privately maintained.

Another instrument which gives interesting readings is a thermometer which is carried on forked twigs close to the surface of the ground in growing grass. It is called the grass minimum thermometer and shows the reduction of the temperature of the lowest layer of air by radiation to a clear sky at night. It has very little to say about cloudy nights.

For those who are interested in determining the motion of clouds an instrument is provided from Sweden with the Greek title nephoscope. The motion can also be followed along a comb of points which can be turned on the vertical rod that carries them so that the cloud appears to travel from point to point of the comb. The instrument was invented by M. Louis Besson. An *Observer's Handbook* explains it.

SELF-RECORDING INSTRUMENTS

In modern practice the task of observing is greatly assisted by the use of self-recording instruments. A circumstance which has contributed not a little to the development of the practice is the invention of aniline ink which will keep a pen writing its record for a week or even for a month. In the earlier days recording instruments required such elaborate management that a formal observatory was necessary to house them. Nowadays an aneroid barometer is made to write its record on a drum in a case that stands easily on a library shelf. A special form of thermometer, a bimetallic spiral or a curved tube filled with alcohol, gives the same service for temperature; and a bundle of hairs stretched between two points to work a lever, by its expansion with dryness or contraction with wetness, keeps a record of humidity. All three together could be got into a cubic foot of space.

The last two want suitable exposure; and all three want regular daily checking by comparison with standard instruments read directly; but as providing a book of the words of the play of weather they are marvellously convenient.

A very valuable recording instrument is the hyetograph, the Greek name for one which records the fall of rain, and incidentally indicates its duration and consequently the rate of fall.

Another attractive instrument is the sunshine recorder (fig. 37). It consists of a glass ball properly turned to be truly spherical, mounted at the centre of a framework in which strips of blue card can be placed to take the image of the sun that appears as a bright spot, so hot that it burns the card and leaves a charred line to tell the tale of sunshine for the day.

Fig. 37. Sunshine recorder.

One of the practical attractions of the instrument is that it excites competition between the "health resorts" of the country for a sunny reputation. The "record" for the day, the month or the year is an avenue to fame; and, for the scientific, the instrument is extraordinarily interesting from the astronomical point of view. A specimen ought certainly to be included in the educational equipment of all schools. By merely noting the position of the spot one ought to be able to read off the time of day—for a sunshine recorder is a perfect sundial—the day of the year—and that, of course, includes the day of the month and the month of the year; and a little ingenuity would give the length of the day in different periods of the year. It is hardly possible to imagine so much information compressed into so little space. It is the irresistible attraction of its working, together with the beauty of its design, that retains the instrument as a standard instrument at meteorological stations, and possibly retards the introduction of an instrument that would record not merely the visibility or invisibility of the sun but the intensity of the radiation that comes in by day and goes out at night.

Fig. 38. Anemometer.

There are other contrivances optical or thermal which will record the existence or non-existence of sunshine. The Jordan instrument which makes a photographic record on sensitised paper is an example of the one; the instrument devised by W. H. Dines, or that by C. F. Marvin of Washington, of the other. But so far as the working of the atmospheric drama is concerned it is essential ultimately to know how much of the solar radiation arrives at the earth's surface and how much is otherwise disposed of. The number of hours during which the ample supply represented by sunshine is kept up is also a matter of interest, very keen interest indeed, but it is not the only one.

At present the only instrument which has been devised to give the intensity of radiation at an ordinary meteorological station is the solar radiation thermometer, a thermometer with a blackened bulb in a glass container from which the air has been exhausted—the black bulb *in vacuo*—and that only gives its maximum temperature reached during the day or its actual temperature whenever the observer thinks fit to read it. Considering that the whole play of wind and weather is staged by the operation of heat received by radiation, something more is required.

PROBLEMS OF THE WIND

Among the recording instruments which aid us in considering the play the anemograph deserves the most careful attention. It records the direction and velocity of movement of the air which passes the vane. It requires a very open space for its exposure and like the weathercock it wants the equivalent of a tall spire to carry its vane. That requirement can be met by providing a small hut to hold the recording part, if the site is free from trees a 30 ft vertical pipe to carry the vane and the connecting mechanism. Our little picture (fig. 38) shows the 100-ft exposure arranged by Mr W. H. Dines at Benson where trees had to be reckoned with.

A modern anemograph is on the pattern originally designed by Dines after the destruction of the Tay Bridge in a gale on 28 December 1879. It records the pressure of the wind on the opening of a tube which is arranged as a vane always to face the wind. The tube that supports the vane provides for the transmission of the pressure to the recording mechanism in the hut below.

We need not enter into details of the construction of the instrument. Unfortunately it is so expensive to buy and so difficult to house properly that we cannot regard it as part of the equipment of a private meteorological station; and yet the information which it gives about the wind is so characteristic that everyone who is interested in the study of weather should take an opportunity of studying some of its records. They are to be found in some museums. A sample is reproduced here (fig. 39). There are two traces, one to show the velocity of the wind and the other its direction.

The first thing to notice about the records is that they are not steady lines like the traces of pressure or temperature but show rapid fluctuations like the weft of a ribbon instead of a sinuous thread. There are corresponding fluctuations in velocity and direction. These fluctuations are regarded as "gusts" and are supposed to be caused by a succession of eddies due to the obstacles in the path of the wind at the surface. There may be as many as twenty such fluctuations in a minute. A man in the street might perhaps be surprised to learn that there might be twenty gusts in a minute; he would reserve the name for some wind of exceptional force that would last more than three seconds. Such things might also be found in the trace; we should attribute them also to eddies larger than the ordinary run. The man in the street may allow us to borrow the name for the smaller eddies too.

We should be very glad if the gustiness could be avoided because the gusts are too irregular to throw much light on the play of wind and weather. They are all play. Weather is associated with the general sweep of the line which can be drawn along the middle of the trace.

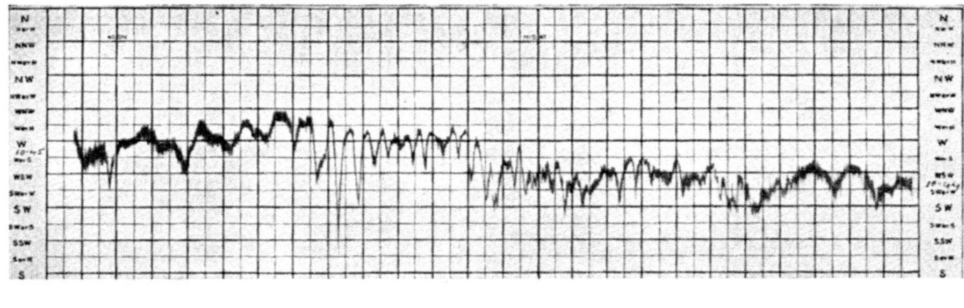

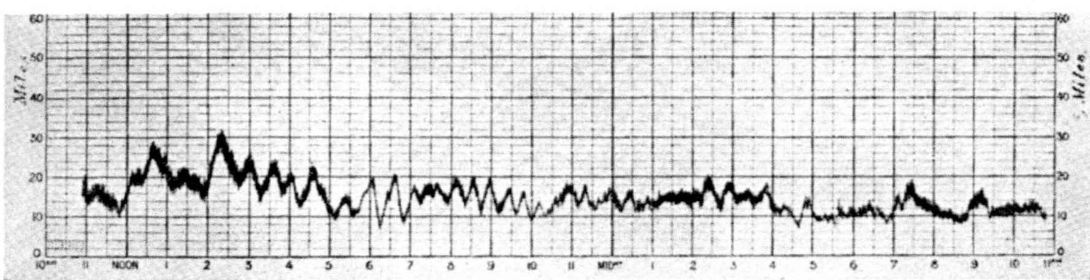

Fig. 39. Wind record, anemogram, from Marshside, Southport, 6 to 7 January, 1907, showing oscillatory changes. Upper panel—direction, lower panel—velocity.

The sweep of the curve gives some indication of the winds that cradle and carry the clouds; but even then the irregularity of the ribbon must not be forgotten; because to produce these fluctuations some of the force of the original wind has been spent. From our point of view it is wasted, and the real and true wind for our purposes can only be reached at a considerable distance above the surface. It is not to be wondered at that a meteorologist watching a sensitive weather-vane feels that the play is not quite fair.

That is one of the reasons for pointing to the upper air. We shall find that for a particular epoch for which a map is made the distribution of pressure shown on the map is a better index of the wind that carries the clouds than the best anemometer-traces.

By the construction of the instrument the traces of the anemograph show the horizontal velocity of the wind. One of the easiest ways of deceiving oneself in watching the play is to think that the wind naturally moves horizontally because at the surface no other general motion is possible than along the surface and all wind-vanes are horizontal; yet it is scarcely unfair to say that the horizontal motion which we watch so carefully is of less importance than its up or down motion which is seldom measured at all.

ATMOSPHERIC POLLUTION

An instrument that comes more nearly within the resources of those who are interested in the weather and the air we breathe under its various changes, is the automatic recorder of the quantity of dirt in the air, designed by Dr J. S. Owens for a committee engaged upon the study of the subject. By its mechanism a measured quantity of air is passed successively through adjacent small areas of filter paper, moved on at fixed intervals by clockwork. A scale of shades enables the observer to estimate the weight of dirty particles in a cubic metre of the air.

In this connexion it should be mentioned that solid particles in the air can be captured and examined by another instrument of Dr Owens, which puts a supply of the particles upon the cover-glass of a microscope. Another instrument, devised many years ago by John Aitken of Falkirk, enables us to count the number of nuclei upon which water is condensed in drops when the pressure of the air is suddenly reduced. Both of these instruments are called dust-counters and dust is a conveniently indefinite term for what the instruments show.

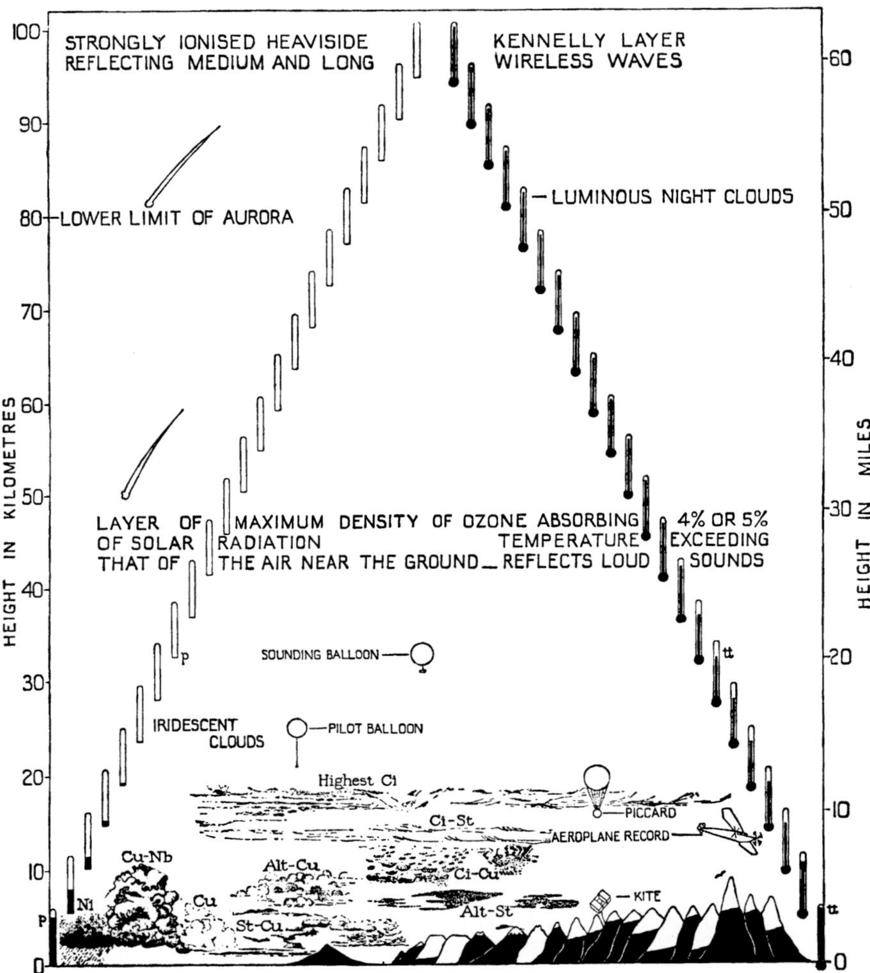

Fig. 40. A pictorial summary of the general results of the exploration of the upper air with suggestions of the various heights observed or inferred and a guide to the corresponding variations in the readings of the barometer and thermometer.

The isolated mountain represents Mt Pico in the Azores and a small mark on the right of it the Eiffel Tower. White mountains suggest perpetual snow and polar regions (see p. 300).

[More recent observations put the layer of maximum density of ozone below 50 km.]

THE EXPLORATION OF THE UPPER AIR

The invitation of the spire pointing to the sky for rectifying the limitation of ideas expressed by a horizontal wind-vane met with a vigorous response at the end of the nineteenth century and the beginning of the twentieth; and in certain directions the activity was more than redoubled as a result of the development of aircraft and the need of a knowledge of the upper air for aerial navigation. And yet from a scientific point of view we may perhaps regret that the necessary activity about the middle layers has to a certain extent damped the ardour of exploration of the range from 8 kilometres to 20 kilometres in height, 5 miles to 12½ miles. Beyond the range of ordinary aircraft, however, individual curiosity has become very active about an ozone layer at 50 kilometres where the air is as warm as that at the earth's surface and echoes the sound of explosions in a very gratifying manner, and a layer at 100 kilometres which reflects electric oscillations. (See fig. 40.)

Beginning with the pressure and temperature of the air, exploration made use of passenger balloons and kites. The British Association reported ascents by James Welsh of Kew Observatory in 1852 and famous ascents were also made for the British Association by James Glaisher in 1862—and therewith began the inquiry into the proper way of obtaining the temperature of air when the thermometer was liable to be exposed to the sun.

The exploration of the upper air by means of manned balloons was developed to its greatest extent by the Aeronautical Society of Berlin which had published three volumes of memoirs in 1900.

The use of kites for scientific purposes goes back to Benjamin Franklin's catching an electric spark-discharge with a kite and the consequent identification of the alarming lightning-flash as an exaggeration of the little spark that can be got from a stick of ebonite when it is rubbed with catskin. Electricity for a spark can also be collected by a wire leading from the wick of a burning spirit-lamp in the open air above any roof to a pair of pith balls hung by a cotton thread in the room beneath. The experiment has still a halo of mystery even in these days of knowledge.

The method of exploration by kites was improved in Australia through the introduction by Hargreaves of what is called a box-kite, a frame of sticks on which fabric can be stretched to form a box-shaped band at the top and another at the bottom. With fine piano-wire to hold it and an electric winch to let the wire out and pull it in again the box-kite can carry an instrument safely to great heights, and it is possible to join on a second kite to carry some of the weight of the wire when the first kite has reached its limit. As many as seven kites have been used in series in that way and heights of 7 or 8 kilometres attained.

The method was employed in 1896 by Abbott Lawrence Rotch at an observatory which he founded for the purpose at Blue Hill, Massachusetts, and about the same time by Léon Teisserenc de Bort in Paris and at an observatory which he founded at Trappes which is not far from Versailles, also by H. Hergesell at Strasbourg and by R. Assmann at the Tempelhofer Feld at Berlin and subsequently at Lindenberg. Kites were used in England by W. H. Dines at Oxshott and at Watlington, and in the United States until recently by the Weather Bureau. They are still in use at Lindenberg and in Brazil; but 6 or 7 miles of wire, with one end only made fast, is a serious responsibility when there are motors on the roads. Teisserenc de Bort related some thrilling experiences of fallen wire engaged with a pleasure boat or a locomotive engine, which were just short of being fatal accidents.

The next stage is the kite-balloon, a curious huge sausage-shaped affair inflated partly with hydrogen and partly with air. That also is manageable by a kite-wire, or a thin cable which is more trustworthy. It is a development of the captive balloon, an ordinary spherical balloon tethered to its base; but a captive balloon, quite good for investigating land-fogs, is only usable when the air is practically calm. When there is anything to be called a wind it becomes quite unwieldy and is very likely to break its tether.

All these devices, kites and kite-balloons and captive balloons, require a self-recording instrument of special design as light as possible and carefully protected from the direct influence of the sun. The record is required to show the pressure of the air, which is quite closely associated with the height, the temperature and the humidity. That is as much as can be expected in the ordinary way.

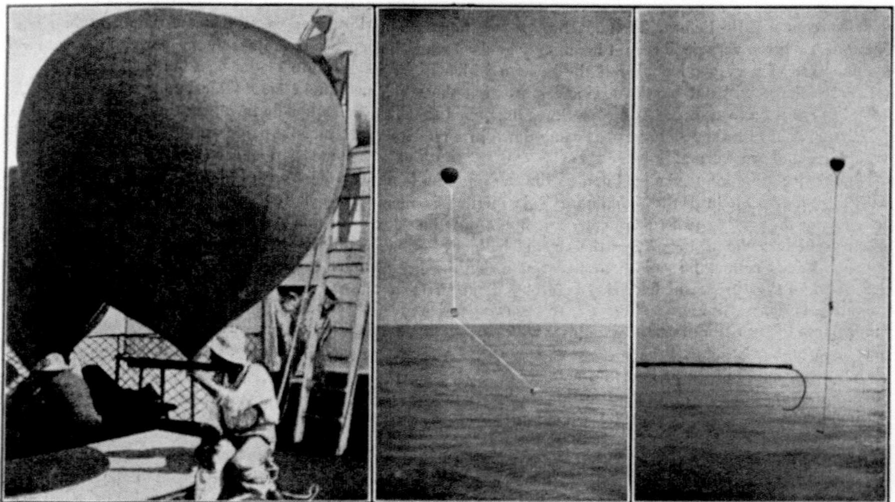

Fig. 41. Launch and recovery of sounding balloons.

From the kite or kite-balloon to the ballon-sonde, an unmanned balloon which carries the recording instrument with the offer of a reward to the finder if he returns it, is the next step. The carrying balloon is arranged to burst when the maximum height is reached, and the instrument comes down, with its gear acting as a parachute, or two balloons are harnessed one of them to burst leaving the other to act as a parachute. There is a vast and only partially explored possibility of the use of this method at sea (fig. 41), a most attractive opportunity for seagoing ships on pleasure-cruises without any rigorous time-table.

These are now in use in many countries for the investigation of the upper air. Special days are noted as international days for ascents, and the combined results are issued as a separate publication. The countries which presented results for ballons-sondes in the publication for 1924 were Austria, Canada, France, Germany, Great Britain, Holland, Italy, Russia, Spain and Sweden. The number has now been considerably increased.

If the weather is clear the journey of each balloon can be tracked by an instrument which will give the distance of the balloon and the angle of its elevation above the horizon.

Such observations are not limited to balloons which carry recording instruments; smaller balloons are sent up in multitudes without any load except perhaps a tail to be observed. Such are called "pilot balloons" because they were originally employed to indicate the track which a passenger balloon might be expected to follow.

Hundreds of pilot balloons are now sent up daily for the information of aircraft. Fifteen thousand are reported as sent up in the British Isles in 1931, and the results form a considerable addition to meteorological literature. Moreover a new form of sounding balloon has been introduced, called a ***Radio-sonde***, which announces successive steps in the change of pressure and temperature of the air through which it travels, by means of a radio-signal produced by electric contact. It was originally introduced in Russia by Moltchanoff and has now been adopted in other countries.

This completes the meteorological equipment of the exploration of the air up to about 20 kilometres. Special apparatus and arrangements are necessary for dealing with the behaviour of the atmosphere in respect of ozone, the sound-reflecting layer and the stories told by meteors.

SOME WEATHER TOYS

We have treated the watching of the drama of the weather as something which may have an interest beyond the circle of the technical expert; which indeed invites the attention of those who can appreciate the arts and crafts of nature without the preliminary discipline of scientific graduation. The idea thus embodied sorts naturally with the fact that various contrivances are to be found on the market, especially about Christmas time, which depend upon the weather, but rank rather as weather toys than as meteorological instruments.

With the extension of the use of scientific apparatus for pleasure as well as duty we may perhaps include the weather-glass with its barometer and its impressive legends, and even the portable barograph, among the scientific toys. Numbers of either are kept and watched by people who would make no claim to scientific training.

Another which, in an expert's view, may be said to share the characteristics of these two, is the Old Dutch weather-glass which consists of a flattened glass globe with a kettle spout issuing from the bottom of the globe and curling upwards like any other kettle spout. Some water is introduced into the globe which can hang on a peg so that the water lies between the air above it in the globe and the open air above it in the spout. In fact a glass kettle with an air-tight lid and partly filled with water would represent the whole contrivance.

With the globe hanging upon the wall we can watch the position of the water in the spout. It will rise if the pressure outside falls and fall if the pressure outside rises, so it is a sensitive barometer, and if it is kept in a room of even temperature, that is the end of its story. But the air sealed in the globe by the water is sensitive to changes of temperature. If it gets warmer it will expand and push up the water-seal in the spout and withdraw it if it cools.

So it would require an accomplished expert to give the equivalent of the changes in water-level in terms of the change in barometer-pressure; but, in so far as it would remind an observer that the behaviour of the atmosphere depends upon changes in the temperature as well as in the pressure of air, it would serve the useful purpose of suggesting the possibility of some numerical combination of pressure and temperature which might be used to indicate the special character of the air in relation to its environment.

An apparatus working on the same principle as the Dutch weather-glass but more elaborately contrived was in use in the middle of the nineteenth century, under the name of the sympiesometer, for measuring the pressure of the air at sea, because the ordinary mercury-barometer was so much disturbed by the motion of the vessel as to be practically unusable. But that difficulty has been got over.

The influence of the varying humidity of the air offers a number of opportunities for weather toys.

Every year the market is provided with flowers, pictures or ribbons which are coloured with chloride of cobalt, a metal not unlike iron in appearance, and remarkable for its coloured salts. The chloride set on in a layer as paint is blue when the air surrounding it is dry, becoming pink as it takes up moisture from the air when it is approaching saturation.

There is a sufficient natural range of humidity in the air to exhibit the change of colours with very striking effect.

And the curious property of material once alive, to which we have already alluded, whereby it appropriates water from the air or surrenders it according to the changes in the relative humidity of the air in which it is set, and alters its shape to correspond therewith, offers also opportunity for the ingenuity of the toy-maker. The Swiss are particularly clever in that way. The house with Jackie and Jeannie appearing in adjacent doorways of a cottage, one on each end of a slip of wood suspended from the cottage roof by a catgut string or a bundle of hairs, represents dryness by the advance of Jeannie with a parasol and wetness by Jackie with his coat and umbrella.

A strip of wood, made up by fixing together face to face slices with the grain of the one across that of the other, curls with moisture for everybody to see how damp the air is.

Perhaps the most instructive of our toys depending on humidity is a contrivance, said to come from America and now excluded by a tariff, which exhibits effectively the behaviour of a wet bulb. The vital part consists of two small glass bulbs one about an inch or 2 centimetres in diameter and the other about one-third of an inch, less than 1 centimetre. The two are connected by a narrow glass tube about one-eighth of an inch in diameter, out of the ends of which the terminal bulbs have been blown. The larger bulb is covered with muslin and is exhibited as the loose head of a fish which serves as a water-container to keep the muslin wet; and for the smaller bulb the fish's tail furnishes support so that the glass globes and the connecting tube rest on the head and tail, the tube sloping slightly downward. The glass is exhausted of air and contains a pellet of ether coloured to make it easily visible.

Exposed to the air, water evaporates from the muslin, the bulb is cooled and ether-vapour in it condensed and the pellet of liquid is forced along towards the mouth. Arrived there it allows the vapour from below to burst through it; the liquid, depressed by that, falls to the sides and clinging there, hardly visible, runs down to the small bulb, collects and reforms as a pellet.

The creature thus represented as thirsty with a varying thirst goes by the name of "humidor" and the rate of evaporation from the head is indicated by the rapidity with which drink follows drink.

Clearly the moisture in the atmosphere is one consideration and equally clearly the rate of flow of the air past the muslin is another. Either statement can be illustrated by tests which a spectator improvises without any difficulty. As one watches the way in which the creature responds to every kind of influence it is impossible not to wonder what the indication of our wet bulbs really means—whether we have really arrived at the terminus of that inquiry.

INTERNATIONAL ORGANISATION

The onlookers of the weather-drama have been recording their experience from the time of the invention of the thermometer and barometer. We give a specimen (fig. 42) of the record suggested by Hooke, the first demonstrator of the Royal Society, together with a picture (fig. 43) of three of the instruments employed. The date is about 1665.

They are now incorporated in an International Organisation of a very orderly character. There are stations of the first order where self-recording instruments are maintained with auxiliary observations to control the accuracy of their adjustments; observatories these are called, and their activity generally includes some experimental work or research about new methods as well as the prescribed routine. We may perhaps regard them as stage boxes which are used not only to keep an eye on the pageant but also to develop new aspects.

Then there are stations of the second order where a complete set of observations in full detail is obtained three times a day at hours which are selected to give a good "mean for the day", and a satisfactory record for comparison with other places which observe at the same "time of day". Stations have been admitted to the second order where observations are made only twice a day; but that is a concession to national frailty and not quite fair to the international organisation.

Then there are what used to be called "telegraphic reporting stations" and may now be classed as synoptic weather stations which observe twice, three times or even twenty-four times each day and not at the

A SCHEME

At one View repreſenting to the Eye the Obſervations of the Weather for a Month.

Dayes of the Month and place of the Sun. Remarkable houſe.	Age and ſign of the Moon at Noon.	The Quarters of the Wind and its ſtrength.	The Degrees of Heat and Cold.	The Degrees of Drineſs and Moyſture.	The Degrees of Preſſure.	The Faces or viſible appearances of the Sky.	The Notableſt Effects.	General Deductions to be made after the ſide is fitted with Obſervations: As,
14 ♊ 12.46 4 8 12 4 8 12	27 ♉ 9. 46. Perigeu.	W 2. 3. 3 1/3 W.S W. 1	9 3/4 12 1/2 16 10 1/8 7 1/2	2 5 2 8 2 9	29 1/10 29 1/8 29 3/8	Clear blew but yellowiſh in the N. E. Clowded toward the S. Checker'd blew.	A great dew. Thunder, far to the South, A very great Tide.	From the laſt Q. of the *Moon* to the Change the Weather was very temperate, but cold for the ſeaſon; the Wind pretty conſtant between N. & W. A little before the laſt great Wind, and till the Wind roſe at its higheſt, the Quick-ſilver continu'd deſcending til it came very low; after wch it began to re aſcend, &c.
15 ♊ 13.40 8 4 6 10	28 ♉ 24.51.	N. W. 3 4 N. 2 1	9 8 1/2 7	28 1/2 2 9 2 10	29 1/16 29	A clear Sky all day, but a little Checker'd at 4. P. M. at Sunſet red and hazy.	Not by much ſo big a Tide as yeſterday. Thunder in the North.	
16 ♊ 14 37 10	N. Moon at 7. 25' A. M. ♊ 10. 8.	S. 1	10	1 10	28 1/2	Overcaſt and very lowring.	No dew upon the ground, but very much upon Marble-ſtones, &c.	
	&c.	&c.	&c.	&c.	&c.	&c.		

Fig. 42. The symbol in the first column is the zodiac sign for May, that in the third for April.

same "time of day" but at the same time of the Greenwich clock. They carry also at least a barograph and often a sunshine recorder. The observations are in every case transmitted immediately by telephone, telegraph or wireless to one of the central establishments for redistribution over the world. It is the directors of these central establishments who constitute the controlling body of the international organisation, so perhaps we should be right in calling a synoptic weather station an official stall in the world's theatre. It has all the responsibility of a station of the second order or even more, and is in direct communication with its surroundings.

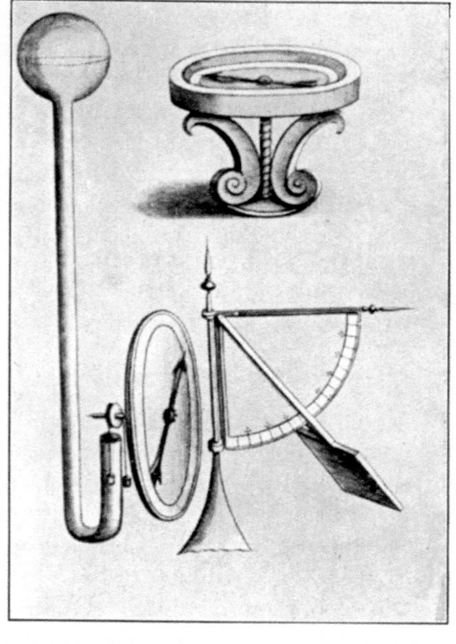

Fig. 43. Meteorological instruments for Hooke's scheme of weather observations.

Besides these there are stations of the third order with less complete outfit of instruments or observing only once a day, an *omnium gatherum* of lookers-on; and a crowded gallery of rainfall stations, perhaps ten for every station of the other orders.

A census of the meteorological auditorium for the world taken by Dr H. N. Dickson for 1914 gave 332 first order stations, 4521 second order, 5877 third order, and 484 unclassified, with a total of 11,214 that can be ranked as climatological. Rainfall stations were not included so perhaps we may assume that rainfall was observed at something like a hundred thousand stations.

Besides stations on land there are ships, five hundred of the British Merchant Navy of which 101 (reduced to 60 for economy) keep weather-watch every four hours of "ship's time", and about 400 out of an allowance of 1000 for the world which take observations for wireless transmission at one or more of the four observing hours fixed by Greenwich time, as well as the ships of the Royal Navy. With the same proportion from other nations there would be 2500 ships to give us information about the weather at sea.

And let us not forget that the atmosphere has height as well as length and breadth, and in that connexion recall that the International Commission for the Exploration of the Upper Air has 102 members from all parts of the world. Most of them are attached to official establishments; but the exploration of the upper air has always had attractions for private persons who are interested in the study of weather.

THE WATCHERS' LIBRARY. BOOKS OF MANY VOLUMES

We come now to the information about the drama that the lookers-on have secured. According to the practice established by the international organisation observations are taken three times a day at normal meteorological stations of the second order, twice a day at others, once a day, and sometimes only a selection of the observations, at stations of the third order, once a day at ordinary rainfall stations, once a week or once a month at remote rainfall stations where rain falls but no one lives. Four times a day at a few stations which report observations by telegraph wire or wireless, three times a day at the great majority of telegraphic reporting stations, twice a day at the reporting stations of the United States and Canada, and also those of the international scheme of Europe. Observations every four hours of ship's time in the ships of the Royal Navy and in 101 ships of the Merchant Navy of Britain with observations at one or more of four observing hours fixed by Greenwich time in some 400 other ships. Corresponding arrangements for the Middle and Far East and the southern hemisphere, and occasional observations by exploring parties in various outlandish parts of the world.

Observations of the wind in the upper air by pilot balloons daily at hundreds of stations, some at sea, by aeroplanes at tens of stations, other observations of the upper air on selected international days. To these add continuous records of wind, pressure, temperature, humidity, rainfall, sunshine, sometimes also solar radiation at stations of the first order or observatories, and miscellaneous observations of solar radiation.

All these are to be found included in the books of the meteorological libraries of the world. The continuous curves are tabulated for hourly readings to be printed *in extenso*, the observations at fixed hours reported in conventional forms and printed *in extenso* or in weekly or monthly and annual summaries. The telegraphic and wireless reports are entered on charts or included in tables on the backs of the charts with reports of clouds, pilot balloons and aeroplanes. The upper air observations for the international days of a year form a volume for the use of students.

One year's observations for the British Meteorological Office alone occupies 3650 large pages of the book for telegraphic reports, 450 pages for observatories and upper air, 200 pages for climatological data weekly or monthly, 300 pages of smaller size for rainfall—allowing some pages for observations from ships at sea which have somehow got lost in what is called "the Hollerith machine" we get a shelf length of a long foot for a single year of British contributions to the library, besides the reserve of original records which may be required for the purposes of study. Anyone who wishes to keep the information in his own library must be prepared to set aside some thirty feet a year; and to make his library anything like complete he must be in communication with officials in every part of the world because the weather-reports of the smaller countries are incorporated with official statistics that do not find their way into the lending libraries.

This goes on year by year with volumes accumulating like water in a reservoir, and has been going on in some form or other for more than 100 years with fragmentary chapters from still earlier times and some occasional information from the early days of civilisation. A volume of the Smithsonian Miscellaneous Collections, No. 79, 1927, contains *World Weather Records*, collected from official sources with monthly values of pressure, temperature and precipitation, extending over many years, Milan from 1764, Berlin 1769, Vienna 1775. It may be regarded as summarising some twenty millions of individual observations.

From these circumstances it results that the book of the play of wind and weather is to be found in full in only two libraries in the kingdom, one at the Meteorological Office and the other nearly next door at the

Royal Meteorological Society. Some indication of it may be found at the local establishment of the Meteorological Office in Edinburgh and perhaps at the observatories at Edinburgh, Dunsink, Oxford and Cambridge; but therein lies a great difficulty in the establishment of schools of meteorology in the British universities—they would first have to be supplied with a book of the play or some substitute for it.

Meteorological science endeavours to abbreviate the extent of its literature by compressing its information into mean values for months or years. So attractive has been the watching of a barometer, thermometer, hygrometer or rain-gauge with a rather diffident interest in the wind that observations have been accumulated in the hope that someone will some day find them useful in explaining the details of the play or the emotions which it excites in the economical world of agriculture, commerce, industry or the health and comfort of mankind.

These have formed what W. J. Humphreys has very appropriately called the frozen assets of the science and it is not uninteresting on occasions to speculate as to whether the scheme of organisation by months arranged according to the "lex divi Augusti" which meteorology stereotyped in the Conferences of Vienna in 1873 and of Rome in 1879 really makes the most palatable preserve of the available material.

Sometimes one is called upon to face a definite question like this—if you wished to keep in mind the emotions which the play of wind and weather had excited in your fruit-trees or your flower-garden, your summer resort or yourself, would you be content with the mean for the day and the mean for the month and the mean for the year? The answer may perhaps become apparent as we trace the action of the drama and realise that weather is a sequence of events and not a registration of mean values. And so when one thinks of the enormous accumulation of frozen assets one may hope that it would somehow be possible to find the actual details of the salient features of the record. This perhaps we might insist upon—sufficient light upon the difference between day and night and between summer and winter, in moisture or the drying-power of the air and in its temperature. And for the rest we must look forward to the time when the real characters of the drama can be identified with stage-directions sufficient to reconstitute the real event.

CHAPTER III

THE SCORE

The use of observation is in noting the coherence of causes and effects, counsels and successes and the proportion and likeness between nature and nature, force and force, action and action, state and state, time past and time present. Exempla illustrant non probant—*examples may make things plain that are proved; but prove not themselves.*

Francis Bacon's letter to the Earl of Rutland.
LORD BIRKENHEAD, *Five Hundred Best Letters*, p. 66

Means, monthly means, how shall we tell your meaning?

THE WATCHER AT HOME

In the previous chapter we have cited the means which are employed to provide material for our memories to re-enact any phase of the weather-drama that we wish to recall in explanation of our experience, or to think about as a scientific exercise. In this one we propose to consider how observation can be used, according to Bacon's suggestion, to put before the reader a summary of the action of the play and to suggest leading motives for the sequence of events in the weather's arena.

First, we shall give examples of the way in which an observer can display the results which he obtains from his own point of view, and next, how his observations can be co-ordinated with those of others for different parts of the world.

The first part of the programme is necessarily chronological. It concerns itself with the sequence of events at a single station and the relation of before and after. The second is geographical, concerning itself with the corresponding observations in other parts of the world.

In that connexion let us note the difficulty under which the science of meteorology labours in having to take into account time and space together, both of equal importance. For the geographer space is the leading consideration, the state of the earth now is the main subject of inquiry, variations in the past or the future add a little zest to the main subject; but in meteorology variations in time are just as much of the essence of the investigation as variations in space.

Self-recording instruments.

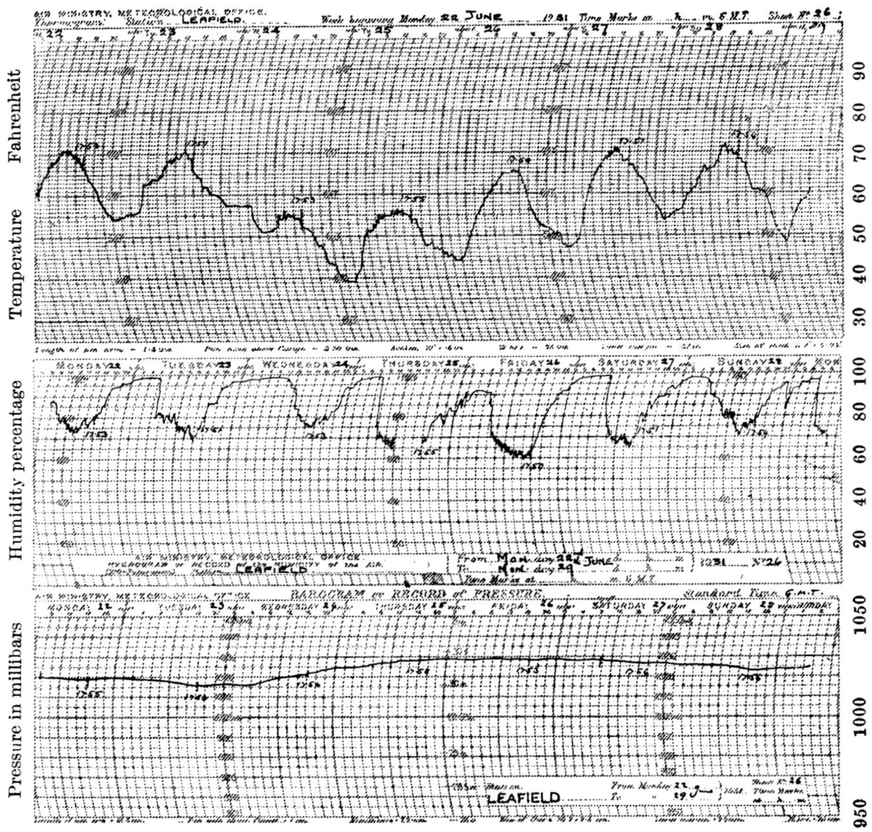

Fig. 44. A week's record (on reduced scale) of temperature, humidity and pressure at Leafield, Oxon, from 22–29 June, 1931.

* * *

Sunshine is also recorded on a card for each day marked by a line scorched by the image of the sun formed by a glass sphere, so long as the rays are not intercepted by cloud.

The accepted practice of meteorological literature is to deal separately with the two aspects, to keep the successive records of a station, or of a series of stations, for the same day, the same month or the same year, and display the results in a series of monthly or annual tables or charts. Sequence is often represented for the separate stations, but when the geographical distribution is exhibited the reader is left to turn over the pages and combine the series in his own way.

We shall exhibit some of the results thus obtained; but the ultimate aim of this chapter is a method of displaying on one chart the two aspects together, chronological as well as geographical.

The difficulty is most serious with observations of wind when we have to take time into account and velocity or force as well as direction.

Let us begin with the single station.

Imagine an observer who is engaged in keeping the record of what he can see at his own station as though he were an occupant of the back seat of a stage box in a theatre, where his view of the stage is limited. To simplify matters we shall imagine him fortunate enough to possess examples of the portable self-recording instruments, now obtainable from any competent maker of instruments, for recording pressure, temperature, humidity and sunshine, as well as the ordinary equipment of a "station of the second order" at which observations are made at fixed hours. These can be used to check the accuracy of the recorders and consequently we shall regard his records as trustworthy.

Autographic records, reduced in scale to bring them within the size of a page, are shown in fig. 44, and as illustrating the remarkable property of hair let us note a similar reproduction in the *Meteorological Magazine* 1933 p. 137 showing agreement of the hair with the wet and dry bulb over the range from 3 per cent to 81 per cent, at Heliopolis.

We shall allow only personal estimates of wind-force because a recording anemometer presents too many difficulties. We shall rely also upon his personal observations of cloud and weather; but we should like him to have a recording rain-gauge because it tells so much that cannot be easily expressed in figures.

As taking part in a climatic scheme for the world we might suppose his hours of observation to be 9 o'clock in the morning, 3 o'clock in the afternoon, and 9 o'clock in the evening, expressed technically as 9h, 15h and 21h local mean time; because those hours were selected by international wisdom as the best for the purpose of a general summary.

Local noon, and therefore local time which is reckoned therefrom, is different at stations on different meridians, one hour slow for every 15 degrees of longitude going westward, twelve hours fast or slow for a point which is just on the east side or west side of the datum line of time-reckoning, 180° east or west of Greenwich.

Our observer might prefer to observe simultaneously with other observers in different parts of the world in order to help in the making of a "synoptic chart" of weather. In that case he would make his observations according to standard Greenwich time and might choose two or more of 1h, 7h, 13h, 18h G.M.T., but the hours might be locally very inconvenient and would not supply so satisfactory a summary of his experience as observations chosen according to local time.

Records made on this plan are published in the weather reports of the Meteorological Office, so we will take a week's observations from that source and give the records of sunshine, temperature, humidity as dew-point, rainfall and pressure, following them with the observations of wind, clouds and weather. In choosing this order we remember that the barometer was chosen as a weather-glass. The reader may like to answer for himself the question as to whether it can foretell the weather, by comparing with the barometric record the observations of wind and weather on the one side, and of temperature and dew-point on the other, so its behaviour is shown in the middle of the frame of observations.

A week's observations and the grand total

The week chosen, or actually eight days, covers the weather at Croydon from midnight on 28 January to 6 o'clock p.m., 18h, on 4 February 1929, a period included within the year April 1928 to March 1929, which for Kew Observatory will be displayed in a subsequent section (p. 196). The week's observations are here extracted from the Daily Weather Report to indicate, when the time comes, the amount of material which can be represented by a suitable diagram and to suggest a method of representation which the ordinary person interested in the weather might use with advantage to fortify the memory of his own experience.

The entries follow well-recognised conventions for reporting the various elements except the direction of the wind; and in that case, for economy of space, the letters **b** to **h** are used, instead of bulky capitals, for the east side of the compass and **t** to **z** for the west side, **a** stands for north, **e** for east, **s** for south and **w** for west, **0** for calm, **c** for NE, **d** for ENE, **f** for ESE and so on. To mark the finer divisions we may note **sh** as S′E and **st** as S′W and so on. The notation for weather is that introduced by Admiral Beaufort. It is reproduced on p. 106.

Cloud-amount is estimated as the number of tenths of the sky covered by the clouds that are visible. A distinction is drawn between the covering due to all cloud together and that which can be attributed to low cloud; because the low cloud speaks in the drama with a peculiar voice.

We have given a very detailed story of a single week of weather, not an easy load for the mind to carry unless it has a store of experience to help it; and if we extend the week to years and allow ourselves to take account of perhaps a thousand official or private "stage-boxes" occupied by fellow-spectators of the weather-drama of the world we can form some vague impression of what the details of the book of the drama really amount to. Allowing for observations three times a day, 25 readings a day, 175 a week, 9000 a year for each station, eight millions a year for the pageant of the weather of the world, we can at least understand that while the details may be necessary on occasions, some method of summarising them is equally necessary.

The agricultural week

The week comes in again to redeem the promise of p. 77 by an alternative method of displaying information about the weather for the use of students of agriculture. Below the table for Croydon on pp. 132–3 are given the salient features of weather at the Cambridge University Farm for the two weeks of April 1933, 16 to 22 and 23 to 29, the 24th and 25th of the daylight cycle described in Chapter VII. Included are the accumulation of sunshine, rainfall and warmth from Nov. 6, the beginning of the cycle, as well as, in black type, the normals for the district, England East, for the corresponding weeks. In Chapter VII the cycle is represented in figs. 105–9.

An official record from the Daily Weather Report of 8 *days demonstrating transitions from cold moist rainy weather to warm rainy weather and then to dry cold weather with sunshine*

		Jan. 28 M.		Jan. 29 T.		Jan. 30 W.		Jan. 31 Th.	
Sunshine hours		0		0		0		1	
Temperature ° F		a.m.	p.m.	a.m.	p.m.	a.m.	p.m.	a.m.	p.m.
on grass: min.		23	—	32	—	41	—	42	—
in screen: min. max.		28	41	34	41	41	50	44	49
Temperature ° F	1h 13h	32	40	36	40	47	47	44	48
	7h 18h	28	39	36	41	47	47	44	45
Dew-point	1h 13h	30	36	34	36	46	44	42	46
	7h 18h	27	37	34	40	46	43	42	44
Humidity relative %	1h 13h	92	85	92	85	97	92	92	92
	7h 18h	97	92	92	97	97	85	92	97
Rain mm (tr trace)	7–18h, 18–7h	tr	3	2	<	tr	tr	<	8
Pressure in millibars	1h 13h	20	19	19	16	20	21	21	18
excess over 1000 mb	7h 18h	19	19	16	17	22	21	19	16
Wind direction	1h 13h	u	s	sh	s	t	sh	sh	h
force		2	4	5	4	3	3	3	4
direction	7h 18h	st	s	s	s	s	sh	h	h
force		3	4	6	5	4	5	4	4
Cloud-amount	1h 13h	10	10	10	10	10	9	10	9
Total	7h 18h	5	9	9	10	10	8	10	9
Low	1h 13h	7	4	10	10	10	9	5	9
	7h 18h	2	5	5	10	10	4	10	9
Beaufort notation	1h 13h	c	c/pr	or	or	or	c/r	c	c
weather	7h 18h	fbc +	c −	cd	or	or	c	c/r	c +

Summary of the record for Cambridge University Farm (See Chapter VII)

Date and number of week of the daylight cycle		Temperature ° F					
		Soil at 9h			Grass	Air	
		4 in.	8 in.	2 ft.	min.	max.	min.
April 16–22 (Week 24)	highest	49·5	50·4	50·0	41	64	42
	lowest	43·0	43·9	47·2	21	45	28
April 23–29 (Week 25)	highest	52·0	50·6	49·8	46	64	50
	lowest	43·2	43·7	46·9	22	54	31

The Croydon observations of sunshine, temperature, humidity, wind, cloud and rain at 1h, 7h, 13h, 18h G.M.T., 28 January to 4 February, 1929

Feb. 1 F.		Feb. 2 S.		Feb. 3 Su.		Feb. 4 M.				
0		0		8		7				Sunshine hours
a.m.	p.m.	a.m.	p.m.	a.m.	p.m.	a.m.	p.m.			Temperature ° F
40	—	40	—	34	—	17	—			min. on grass
43	51	41	45	35	39	25	44			min. max. in screen
45	51	47	45	39	40	30	39	1h	13h	Temperature
46	49	42	43	35	31	25	34	7h	18h	° F at fixed hours
44	47	46	44	35	25	23	28	1h	13h	Dew-point
45	45	40	41	28	21	23	30	7h	18h	
97	85	97	97	85	55	75	65	1h	13h	Relative humidity
97	85	92	92	75	65	92	85	7h	18h	%
2	tr	5	tr	0	0	0	0	7–18h,	18–7h	Rain mm, < less than 1
11	8	14	11	13	15	20	24	1h	13h	Pressure in millibars
7	11	12	11	14	18	23	25	7h	18h	excess over 1000 mb
gf	t	st	h	f	ef	g	y	1h	13h	Wind direction
4	4	2	5	4	4	1	1			force
h	st	g	gh	g	gf	0	v	7h	18h	direction
5	4	4	5	4	3	0	1			force
10	10	10	10	10	1	0	0	1h	13h	Cloud-amount Total
10	10	9	10	9	1	0	0	7h	18h	
10	10	10	10	10	0	0	0	1h	13h	Low
10	10	9	9	9	0	0	0	7h	18h	
or	c	o	or	c	b	b	ffb	1h	13h	Beaufort notation
oo	c	c –	c/r	c	b	b	mb –	7h	18h	weather

for the 24th and 25th weeks of the Daylight Cycle of the Year 1932–33.

	Accumulated Temperature (explained on p. 143)				Rainfall		Sunshine		
	above 42° F		below 42° F		inches		hours		Black figures are normals for England East
Aggregate to April 15	**398**	541	**713**	519	**10·0**	6·27	**420**	553·7	
Current week		24		22		·14		37·1	
Aggregate to April 22	**440**	565	**725**	541	**10·3**	6·41	**459**	590·8	
Current week		65		4		1·25		21·9	

By way of summary

The common method of summarising is by the process of taking what is called a "mean" or average value of the seven observations in the week, at each fixed hour, adding them up and dividing by seven to give the week's mean *for that hour*, and continuing the same process for the days of a month to get the month's mean for that hour, or for the 365 or 366 days of the year to get the year's mean, and still further to get the mean for a period of years to give what is called a "normal" for the specific hour. Summarising further by regarding the average of the three observations in a day as the mean *for the day*, we can proceed in like manner to obtain a mean for the whole week, the whole month, the whole year, or the whole period of the normal. Thirty-five years is a favourite period for normals.

Those fortunate persons who possess recording instruments could read the values indicated on the curves for each of the twenty-four hours of the day and obtain a mean for each of the twenty-four, and so for the day, the week, the month, the year and the period of the normal, a mean which would be more rigorously "correct" than a value depending only on observations at the three hours. But the three hours 9h, 15h, 21h were chosen to give as near an approximation as possible to the results derivable from the combination of twenty-four hours.

We are faced here with the question set out in the heading to this chapter as to what "means" mean. Sir Norman Lockyer used to quote an opinion for which he gave no other authority than his own, "la méthode des moyennes, c'est le vrai moyen de ne jamais connaître le vrai" and there is certainly a substratum of truth in the epigram. In a mean value all the salient features of the original observations are suppressed; the mean temperature for the day is not by any means the most interesting feature of the day's weather, and every other kind of mean is open to similar criticism. For some purposes extremes are more useful than means. A student who has the extremes can make a moderate guess at the mean; but one who has only the mean gets no idea of the extremes which are often of vital importance.

For physical explanation of the process by which the play of wind and weather is carried on, means are merely a preliminary survey; physical science must deal with actual facts. A certain pressure may be appropriate to a certain distribution of temperature, moisture and wind; but the mean pressure does not necessarily correspond with the mean distribution of temperature, moisture and wind. Means are sadly disappointing for those who wish to trace the relation between cause and effect in weather for any of the applications of the science, in forecasting for transport, or in agriculture. But we may, at least, point out that they are specially useful for the self-contained climatology that compares one station with another. If we wish to form a mental picture of the difference of the weather between day and night, between summer and winter, between this year and that year, mean values hourly, monthly, annual come in very conveniently because they form a basis of comparison.

Thus means, hourly, daily, weekly, monthly, annual, are the means by which we make comparison of the behaviour of the atmosphere in different places (geographical) and at different times (chronological). In ordinary meteorology they are, as a matter of fact, aggregates and it is the aggregate which we compare.

Those who regard the collection of observations as material for reasoning on statistical lines use mean values as datum lines for the study of the deviations from the mean. They, too, have other means of summarising the collection, as, for example, by counting the frequency of occurrences of specified values. In the case of winds, for instance, the compilation of observations in simple "means" would conceal important features, and it is natural instead to set out, as a percentage of the whole number of observations, those of calms and of gales, strong winds, light winds, from each of a number of directions. We thus get a series of numbers suitable for expressing our results.

Some persons have the faculty of reading columns and rows of numbers and obtaining general ideas from them, and others find satisfaction only in associating the numbers in an algebraical formula. Some, on the other hand, find more scope for imagination in a pictorial representation of the facts. With these we will take the liberty of associating the reader because the traces which we have already given suggest it. A diagram written by nature herself may almost claim the right to be regarded as a pictorial representation of the score of one of the instruments of a natural symphony.

In the practice of weather-study there is great variety of pictorial representation and we would now direct the reader's attention to a selection.

If we had unlimited facilities at our disposal we should begin by exhibiting a record of the amount of energy received by day as solar radiation and the corresponding loss by terrestrial radiation; but records of that kind are comparatively scarce and as a rule their effects upon weather are not directly indicated. The commoner record of the sunshine recorder is more tantalising than helpful in this connexion, because though the scorched line on the card reminds us of the existence of the energy required for the maintenance of life on the earth, it shows only the duration of sunshine strong enough to char the paper and that is only a small part of the earth's life-history. We are on firmer ground with the representation of the effects, direct or indirect, of the sun's influence upon the atmosphere.

GRAPHS AND POLYGRAPHS

For a sequence of changes the most common pictorial representation is what, in places where they teach, is known as a graph. The actual records of pressure, temperature and humidity, of which a week's specimens have been reproduced (fig. 44), can be included in the general term graph; but they have a peculiarity in having as time-marks, parts of circles along which the pen would move if the drum were not going. With the picture before him the reader will be able to understand how the continuous record gives him the pressure, the temperature or the humidity at any time that he likes to call for it.

Here, as a curiosity in the way of graphs, we include a specimen (fig. 45) of a record of temperature of an unusual character and with an exposure indoors that would exclude it from the meteorological score of the watcher at home. Yet it has points which may interest the most meticulous meteorologist. First, it has obviously the curved lines for the track of its pen, but instead of requiring twelve of the spaces between the

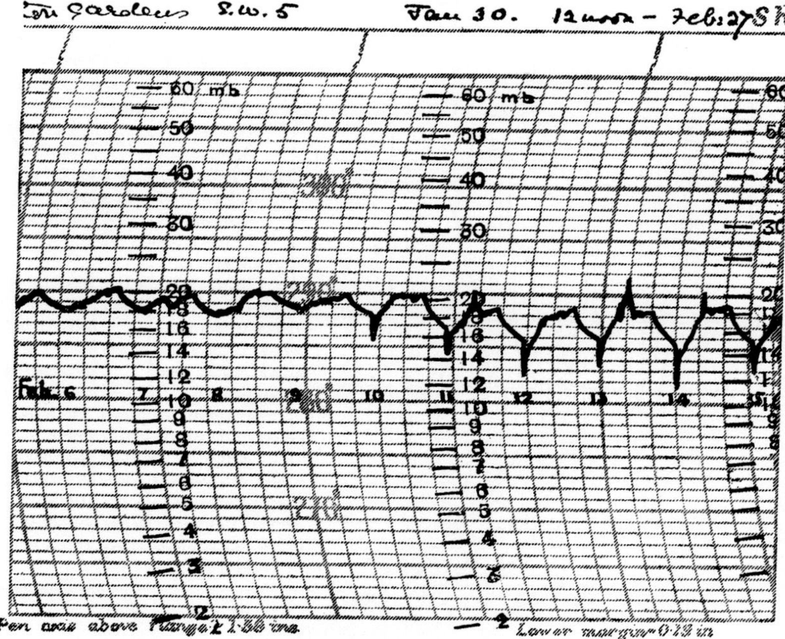

Fig. 45. Ten days' record of temperature indoors from 18h 5 Feb. to 18h 15 Feb. 1933, full size, compressed by the slowness of the clock's motion to show 28 days' record in the space usually occupied by seven.

lines for a day's record it uses only three; each step of one-third of a centimetre is eight hours, a centimetre to a day. The drum which carried it rotates once in 28 days instead of 7 days; and so the picture 10 cm wide shows the variations of temperature in the ten days from 5 to 15 February when the cold spell of the 20th to the 25th which is referred to in Chapter v was in the meteorological offing.

Next, being merely on trial indoors in a living room with ample windows, the cold air of February made itself felt if the windows were open, and from the record we may conclude that within the eleven days there was a change of domestic habit; the windows must have been opened at

Fig. 46. Polygraph of changes of temperature with height above sea-level obtained from balloon-ascents 1907–8.

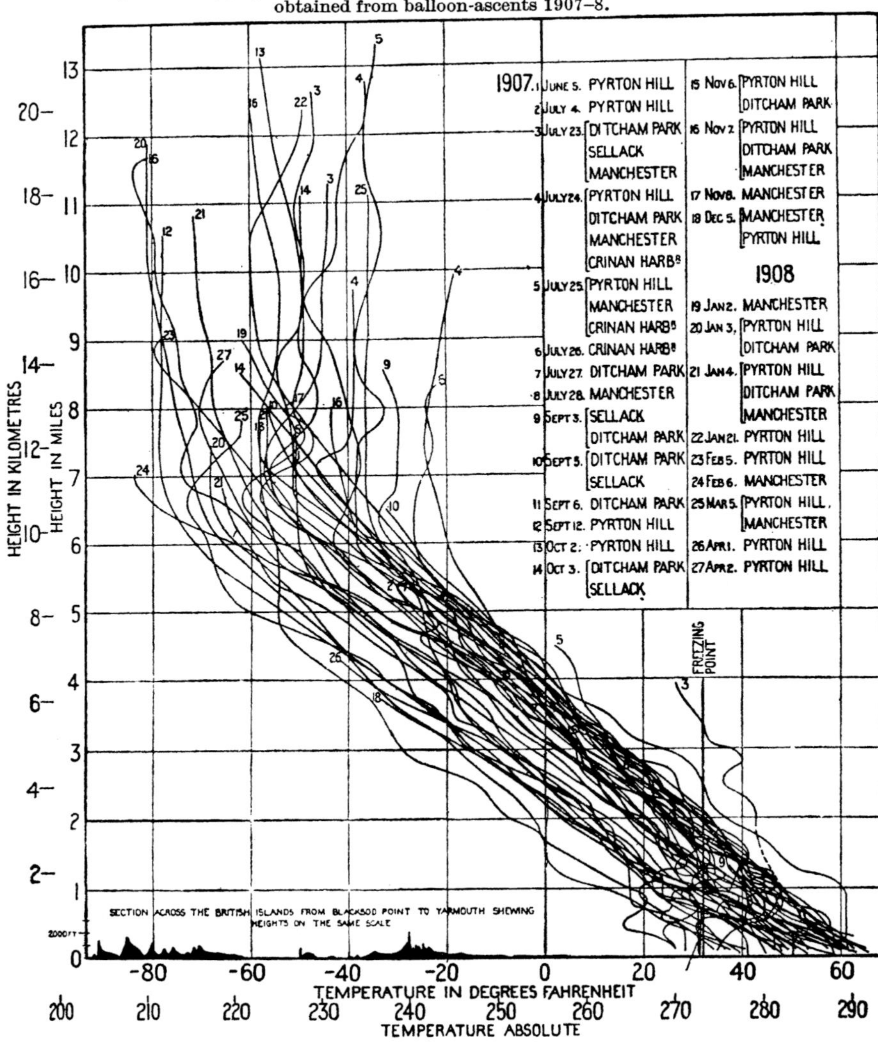

The separate curves represent the relation between temperature, in degrees Fahrenheit, or on the absolute Centigrade scale, and height in miles or kilometres in the atmosphere. The numbers marking the separate curves indicate the date of ascent at the various stations as shown in the tabular columns. The general aspect of the curves shows the great complexity of the temperature variations within the first two miles from the surface, and a very nearly uniform rate of fall of temperature above the two-mile limit until the isothermal layer of the stratosphere is reached, at from six to eight miles. The difference of height at which the isothermal layer is reached, and the difference of its temperature for different days or for different localities is also shown on the diagram by the courses of the lines.

or about 8 o'clock, 8h, on the 10th and following days but not on the previous days.

Thirdly, if the day were fine the sun's rays might reach the instrument through the window in the afternoon and cause an increase of temperature against the like of which special precautions are taken by the competent meteorologist. But the direct sunshine in February is not as effective in leaving its trace as the open window.

And finally a rubber stamp has been employed to superprint on the lines of the temperature scale the maximum of pressure of water-vapour that corresponds with the temperature as set out on p. 104. Objection may easily be taken that the run of the rubber stamp is a misfit; but it is the first of its kind and, once tried, the misfit is an invitation to the reader to do better.

Our next specimen of the graph type, fig. 46, is often used to represent the result of observations of temperature in the upper air. Distances along the base line, or parallel to it, represent temperature, and distances drawn upwards at right angles to the base line show the height at which the particular temperature was indicated on the record. In the diagram, however, which we reproduce, are many graphs, so many that it is not possible to trace the graph for any particular ascent with certainty. But if we may excuse ourselves by calling a group of graphs in a single diagram a polygraph, we may add that since the diagram was originally constructed, it has been found effective on many occasions for illustrating the general fall of temperature with height in the upper air, represented by the slope of the curve from right to left as one passes upward, until in each case a limit is reached beyond which no further fall occurs and the line of the graph becomes approximately vertical.

We call the lower layer in which the temperature falls with singular regularity, the troposphere, and the upper layer in which no further fall occurs, the stratosphere, using names chosen some thirty years ago by M. L. Teisserenc de Bort, who devoted his life to investigations of that kind. The point of change at the junction of the two where troposphere and stratosphere meet, and consequently the region where the troposphere ends, received the name tropopause during the War. The reader will easily see that the tropopause shows a good deal of variation both in its shape—sometimes a gradual bend, sometimes a rapid one—and also in the height at which the vertical run is attained. Otto Pettersen has found something like it in the ocean. It is obviously one of the things which loses its effective significance if we try to express it in mean values.

Temperature and rainfall round the globe

Along latitude 50° N

Along the equator

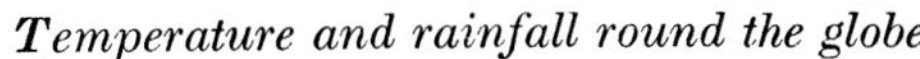

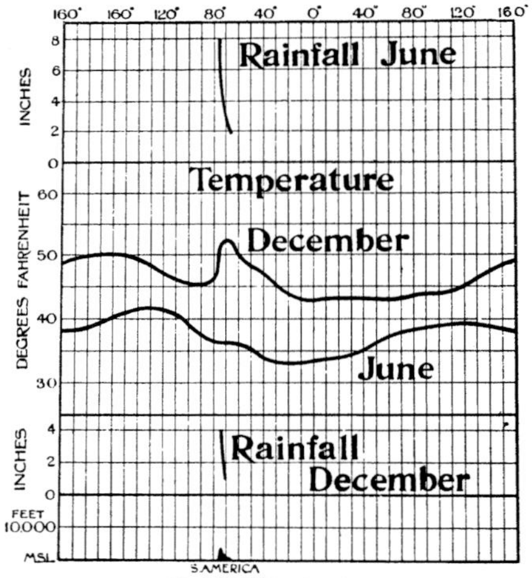

The figures at the head of the sections indicate the longitude from 160° E, Eastwards through 180° to 160° E.

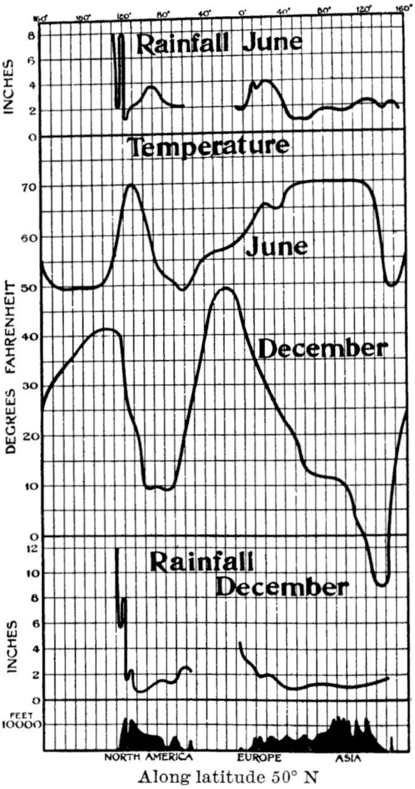

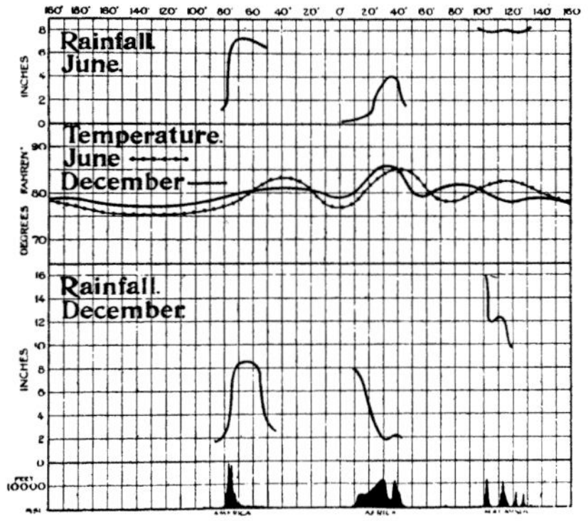

The scales at the side show temperature in degrees Fahrenheit and rainfall in inches.

Along latitude 50° S

Fig. 47. The normal influence of orographic features on weather and climate.

So far our pictures have represented individual observations "themselves alone"; but pictures are also used to express the result of co-ordinating observations by using mean values for days or weeks, or months or years.

Let us take as an example (fig. 47) a graph in which a horizontal line represents distance along a line of latitude on the earth, and use it in order to exhibit the normal temperature and rainfall at different places on the same latitude for selected months. Sections along three latitudes are shown, viz. 50° N, the equator and 50° S. Separate lines indicate the normal rainfall and normal temperature for June and for December at different points along those latitudes. A little black graph at the bottom shows a miniature profile of the land surface above sea-level along the line of latitude to which the diagram refers, with a horizontal line to show the altitude of 10,000 ft.

The main features of the picture are the high temperatures over the sea and low temperatures over the land in December and the reverse in June along lat. 50° in the northern hemisphere: the opposite in the southern. But the fluctuations in rainfall near the West Coast of America in latitude 50° N are worthy of our notice. There we find heavy rain on a coastal range, then a line of much less rain and then again heavy rain over the main range of the Rocky Mountains. Complication of the same kind would be shown where the graph of rainfall for June in latitude 23° N passes over the eastern ranges of Asia.

The reader can hardly fail to notice that while our information about the temperature of the air has something to say about every point on the lines of latitude, there are gaps in the lines for rainfall corresponding with gaps in the profile of the land. There is no corresponding information about rainfall over the sea and he may thereby mark for himself one of the most melancholy features of the present state of the science of meteorology.

Weekly progress in the British Isles

Our next effort exhibits specimens of diagrams (fig. 48) which summarise the observations of temperature, rainfall and sunshine in the several districts of the British Isles as represented in the Weekly Weather Reports of the Meteorological Office for the years 1881–1905. The diagrams are in the form of graphs, using a time-scale of successive weeks in the year instead of a scale of longitude round the globe. They are based upon successive steps in mean values or aggregates for the week, obtained from

Weekly weather in the British Isles.

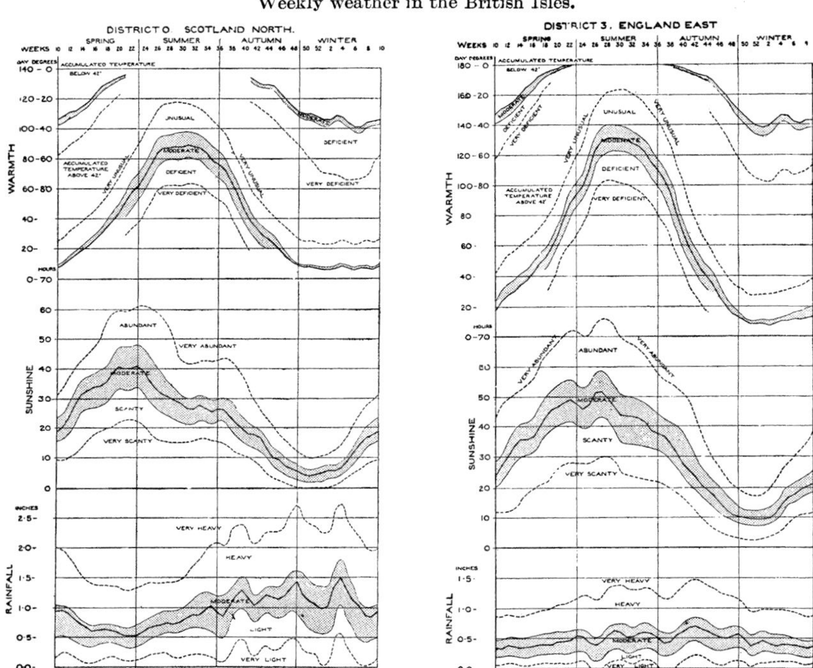

Fig. 48. Curves of normal weekly values in two districts of the British Isles: Scotland North and England East. (See also Figs. 108–9.)

the values of temperature, rainfall and sun, week by week in two of the districts, namely Scotland North and England East as representing notable differences of climate.

As graphs they are peculiar; first because they try to give the reader more information than is contained in the simple mean value and secondly because the records of temperature are treated in a peculiar manner.

In each of the two diagrams rainfall is dealt with in the lowest group which has a scale of inches for each week at the side. Sunshine is shown in the middle group with a scale of hours for each week. Each of the groups includes five lines. The middle line which is continuous and the boldest of the five shows the mean rainfall for each week in inches, or the mean duration of sunshine for each week in hours. Above and below the line of simple means are continuous lines with the space between them shaded. The higher one shows a rainfall or sunshine above the average for the week which is likely to occur once in three years and the lower line indicates the weekly rainfall or sunshine below the average for the week which may be expected also to occur once in three years. The other two lines are dotted and a long way from the mean; they represent conditions for the weeks which may be expected once in every twelve years.

The year represented begins with spring in the 10th week of the calendar year passing into summer in the 23rd week, into autumn in the 36th week and into winter in the 49th week.

The general view of the rainfall graphs suggests for England East a dry spell in the 26th week, the end of June, and in the 38th week, the second week of September (a very notable characteristic), also a wet spell in October, the middle of autumn; whereas for Scotland North the conspicuous features are a spring generally dry, and autumn and winter notoriously wet with three wet spells culminating in the fourth week of the New Year, the end of January.

In sunshine for England East there is nothing very special to notice. For Scotland North the prolongation of the conditions of late summer into the early autumn may have something to do with grouse-shooting on 12 August.

The group at the top of the diagram is devoted to warmth and its opposite, and the scales at the side show for each week the "accumulated temperature" above 42° F with the scale running upwards and below 42° F with the scale running downward from the top.

This is a special view of the effect of temperature which was introduced for the use of agriculture or perhaps for biology generally. It is based upon the idea of a celebrated botanist de Candolle that vegetation only "grows" when the temperature of the air is above 42° F; below that temperature he regarded growth as in suspense until warmer conditions came or the plant might perish with cold.

Whether that be true or not it is an undoubted fact that high temperatures generally promote rapid growth in vegetation and low temperatures retard it. So the accumulation of temperature in "day degrees above 42° F" or below 42° F, as the case may be, has been regarded as an interesting expression which would enable observers to specify the conditions of climate necessary for the growth of different kinds of vegetation.

Day degrees are easy enough to understand and compute if the temperature is above 42° F or below that figure consistently all day and all night; they are then represented by the excess of the mean temperature above 42° or below it; but for days which have a maximum above and a minimum below the critical figure a formula is necessary and was provided by Sir Richard Strachey. Punctilious accuracy is not really worth much concern if it is the accumulation that is used for comparison, because unless the day temperatures are high a day's addition to the accumulation is not important.

With this preface we may add that the scale at the side shows the weekly steps of the accumulation with the line of mean value as before and the range for a probability of one year in three and one in twelve.

At the very top are the curves for accumulated temperature below 42° F. They are not very impressive but we may remark that those for England East show a little more activity than Scotland North and that also means something for the lives of plants.

PICTURES BY LINES OF EQUALITY DISPLAYING NORMALS FOR HOURS AND MONTHS

We have already alluded to the practice of forming means for each hour of the day and grouping the figures for the hours according to months and so getting the monthly values of means for each of the 24 hours. These can be put together and made the basis of very instructive diagrams which carry the uncommon name of isopleths, being composed of lines indicating the same magnitude at every point of their run. They differ from the scholastic graph in that the lines are drawn to mark the equality of the measure of some element at a selected station—of pressure, of temperature, of humidity, of duration of sunshine or of wind velocity.

Temperature

Let us look, for example, at the table of normals for 25 years of temperature at Kew Observatory, which sets out the normal temperature at midnight, 3h, 6h, 9h, noon, 15h, 18h, 21h and again midnight, all according to Greenwich Mean Time for the twelve months of the year. The use of Summer Time would put the figures for the summer months an hour out of gear.

Normal temperatures at intervals of three hours in the several months of the year at Kew Observatory, Richmond, based on observations for the 25 years 1871–95 *in the Fahrenheit scale.*

Hours	Midt.	3	6	9	Noon	15	18	21	Midt.
January	*37*	*37*	*37*	*37*	40	41	*39*	*38*	*37*
February	*38*	*38*	*37*	*38*	42	43	41	*39*	*38*
March	*39*	*38*	*38*	41	46	47	45	41	*39*
April	43	42	41	47	52	53	51	46	43
May	48	46	47	53	57	59	57	51	48
June	54	52	53	59	**63**	**66**	**63**	58	54
July	58	56	57	**62**	**66**	**68**	**67**	**61**	58
August	57	56	56	**62**	**66**	**68**	**66**	**60**	57
September	53	52	51	57	**62**	**63**	**60**	55	53
October	47	46	45	48	53	54	50	49	46
November	43	42	42	43	46	47	44	43	42
December	*38*	*38*	*38*	*38*	41	41	40	*39*	*38*
January	*37*	*37*	*37*	*37*	40	41	*39*	*38*	*37*

Black type for 60° F or above, italic below 40°.

The original figures from which this table is derived are used again in Chapter IV to represent the diurnal rhythm in temperature for each month in fig. 70.

For simplicity's sake the figures are rounded off to show whole degrees Fahrenheit. It is easy to see that the prize for the warmest hour is divided between July and August, for their performance at 15h (4 o'clock summer time), and there is little or, in our table, nothing to choose between a number of claims for the coldest hour. If we go into decimals 6 o'clock in the morning of January just gains the mark of distinction. We can evidently draw a line separating the sixties from the fifties and the fifties from the forties, and the forties from the thirties, and thus divide the field representing the year by lines which guide the eye to summarising the whole year's normal history as regards temperature. In the printed table the grouping is indicated by using black type for temperatures of 60° F or more and italic type for temperatures below 40° F.

The same sort of diagram would be arrived at with observations every four hours or every six hours. It gives the most concise representation of the daily and seasonal rhythm of an element at a single station.

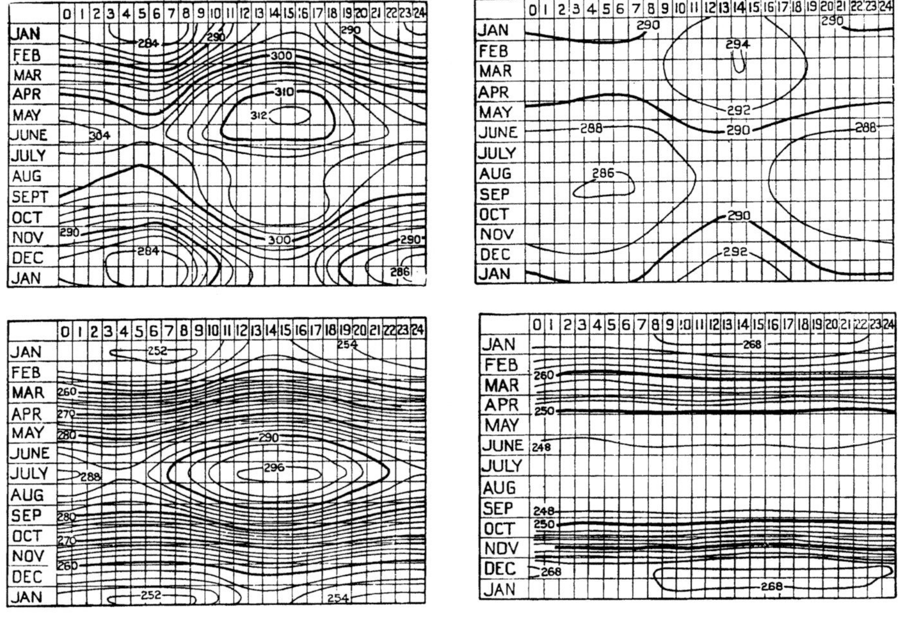

Fig. 49. The upper pair represent the diurnal and seasonal variation of temperature at Agra lat. 27° N long. 78° E on the left, and St Helena lat. 16° S, long. 6° W on the right; the lower pair corresponding information for Barnaul lat. 52° N, long. 84° E on the left, and the shore of the Antarctic continent about lat. 77° S on the right.

More workmanlike examples of the isopleth diagram are shown in fig. 49 to illustrate the daily and monthly regime of temperature at Agra in continental India, at St Helena the oceanic island in the South Atlantic, at Barnaul in Siberia, and at a combination of stations in the Antarctic. Only the lines for selected temperatures are drawn in the diagrams, the actual figures for temperatures do not appear and the scale of temperature used is not that of Fahrenheit, nor even Centigrade, but that which we have called tercentesimal and which is arrived at by adding 273 to the Centigrade reading. That makes no difference to the run of the lines. The line marked 310 corresponds with 98·6° F, 300 with 80·6°, 290 with 62·6, 280 with 44·6, 270 with 26·6° F, 250 with 9·4 below the Fahrenheit zero − 9·4° F.

At Agra the day-temperature reaches its maximum, greater than 312 tt (102° F), at 3 o'clock in the afternoon of May, and the night-temperature its minimum between 5 and 6 o'clock in the morning at the very beginning of the year, showing a range from less than 284 to greater than 312, that is 28 tt (50° F); whereas at St Helena, an island in the South Atlantic Ocean, we find only a range from 286 tt (55° F) at 5 a.m. when August changes to September, to 294 tt (70° F) at 2 o'clock in the afternoon when February changes to March; the whole range for the year being of the order of only 8 tt (15° F).

In the matter of contrast Agra is out-done by temperature at Barnaul in Siberia, where the change is from 252 tt at 6 a.m. in January to 296 tt between 2 and 3 o'clock in July, a range of 44 tt (79° F). In contrast again we have the diagram made for stations in the Antarctic, where there is considerable seasonal change between summer (January) and winter (June, July, August); but hardly any change in the 24 hours. That only shows in the lines about December and January—high summer in the Antarctic.

Sunshine

In like manner we may exhibit isopleth diagrams of sunshine. For this purpose we have to collect suitable information from a sunshine-recorder exposed at the station for which the diagram is required. The information usually given about sunshine states the total number of hours covered by the trace scorched on the card of the recorder and for the isopleths we require to know the behaviour of the sun at each hour of the day when the sun was above the horizon.

Hence the card must be tabulated to show what fraction of each hour was covered by the record and when the figures for the same

Fig. 50 a. Sunshine in various localities.

Aberdeen 57° N

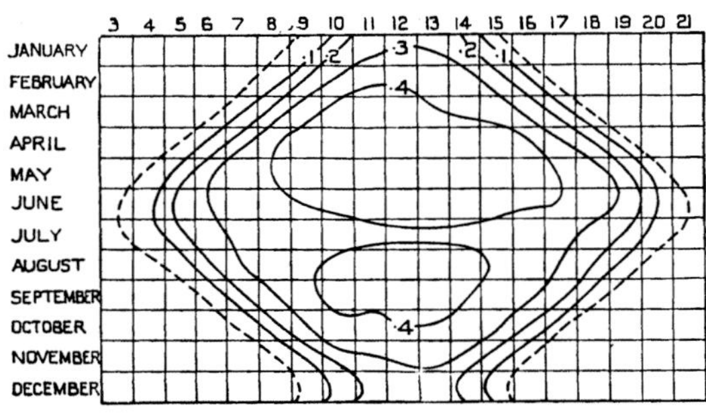

Batavia 6° S

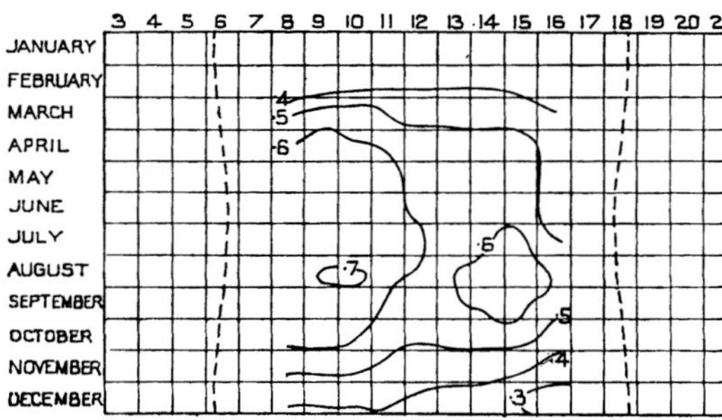

Falmouth Observatory 50° N

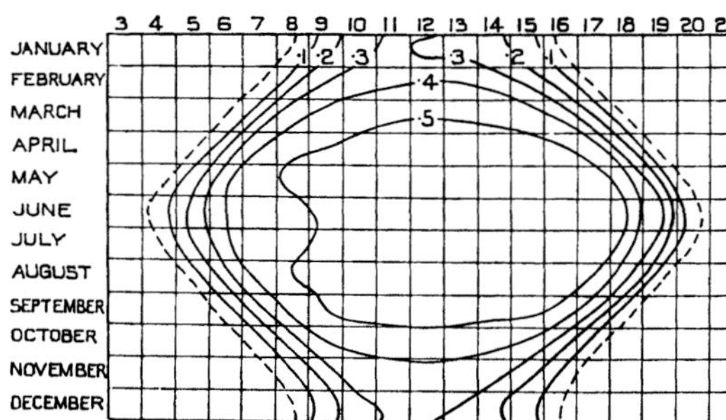

hour of each day are known the total will give the numbers of hours of record obtained within the limits of the selected hour, and the mean for all the days included in the total will give the daily fraction of the hour devoted to sunshine. Of course the total may be made up of all the fractions between 0 and 1 and must include the days when the sun shone throughout the hour as well as the days when it did not shine at all and the days when it shone only for a few minutes, enough to mark the card. In means we take the thick with the thin.

Six examples of such isopleths are reproduced in fig. 50, to show the general character of the hours at Falmouth in the south-west of England, at Victoria on Vancouver Island in about the same latitude off the west coast of North America, at Batavia in the Dutch

East Indies and at Georgetown, British Guiana on the north coast of South America, both near the equator, at Aberdeen in the northern hemisphere and Laurie Island at nearly the corresponding latitude in the southern. The north exhibits itself to great advantage. In these diagrams the figuring ·9 in Georgetown tells us that the hour between **10.30** and **11.30** had sunshine for nine of the hours out of every ten, ·4 means the equivalent of four sunny hours out of ten and so on for each of the diagrams.

In each of them there must always be a period of night, when the station is in the earth's own shadow; so a dotted line on the right hand side "tolls the knell of parting day" and a corresponding line on the left suggests on the other hand the rosy-fingered morn leading on to "the warm precincts of the cheerful day".

Fig. 50*b*. Sunshine in various localities.

Laurie Island, South Orkneys, 61° S

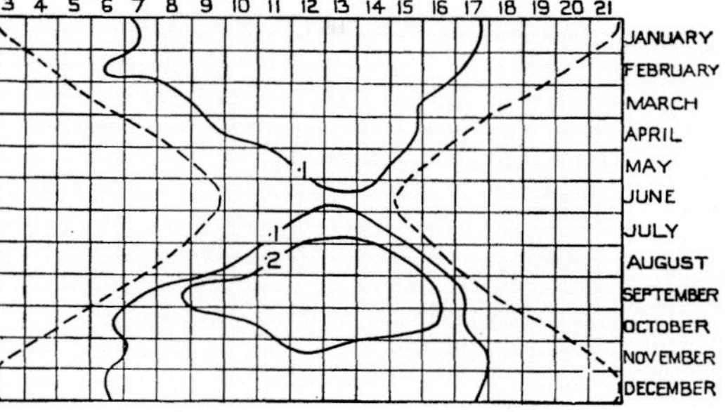

Georgetown, British Guiana, 7° N

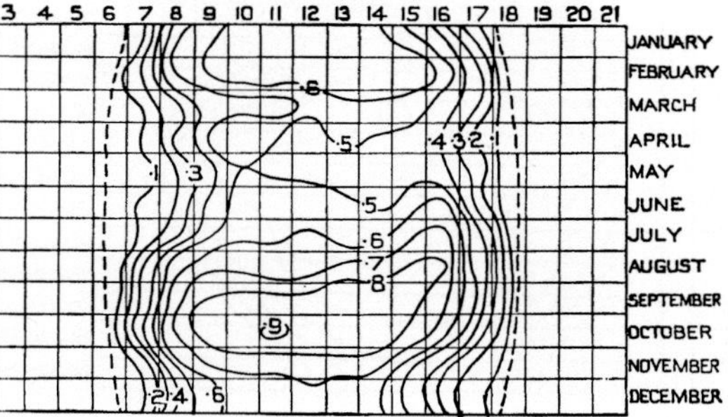

Victoria, Vancouver Is., 48° N

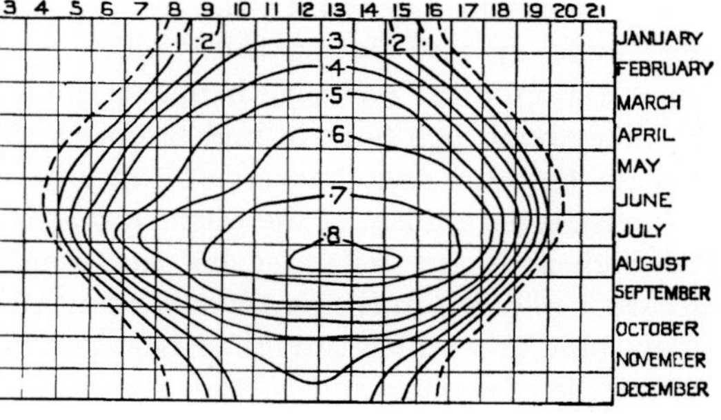

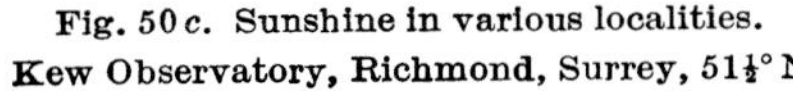

Fig. 50 *c*. Sunshine in various localities.
Kew Observatory, Richmond, Surrey, 51½° N

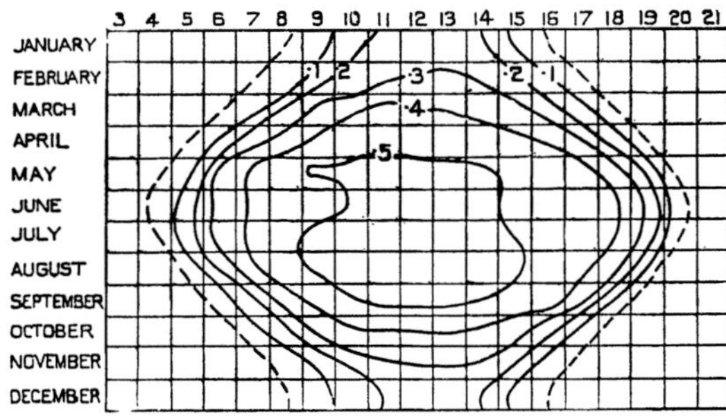

Valentia Observatory, Co. Kerry, 52° N

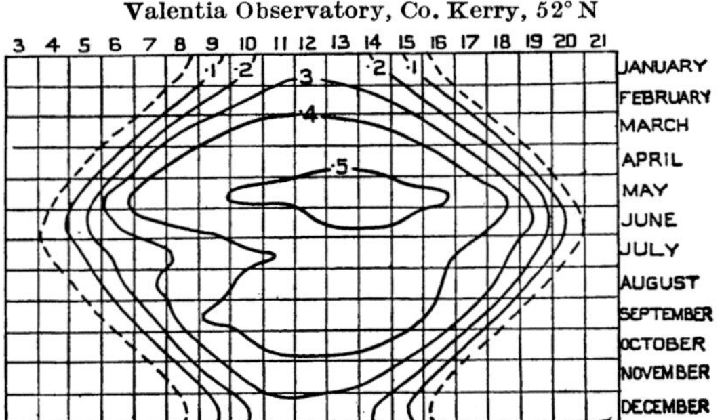

Two additional diagrams enable us to compare the records of sunshine at Kew Observatory in the east of England with Valentia Observatory on nearly the same latitude in the south-west of Ireland. The best of hours at either can call only half its duration sunshine. The month of May shows a prominence of sun in the early hours in both. Kew gives to July and August such preference as it has at its disposal, whilst Valentia makes a recovery in September from July's notable poverty.

Sunshine and temperature lend themselves easily to representation of their general features by diagrams like those that we have presented in the pages which precede. Running along the line of hours in any month the middle figure is for mid-day, and for any hour, up and down, the middle figures are for mid-summer. For a station in the northern hemisphere the central portion represents the time about high noon at midsummer and it is not surprising that it displays the abundant harvest of sunshine appropriate to those hours and, with a certain lag, of the warmth that is derived from the sunshine. For high southern latitudes on the other hand, the centre of the picture is mid-day in the depth of winter. All the curves of equal amount of sunshine or warmth turn their backs on it and display abundance in the middle of the top and bottom margins.

Rainfall

Rainfall must also be the result of the warmth due to sunshine; but we do not generally recognise it as directly dependent upon the sun's position in the sky or in the zodiac. To form an opinion about that we can put the figures for rainfall into a diagram just as easily as those for sunshine or temperature and they should enable us to judge whether there is any direct association of rain with the sun's position and, if so, of what kind.

It is quite likely that in the regions not too far from the equator, where the will of the weather gods is ordinarily expressed in thunderstorms, a diagram of rainfall or rain-hours would show a marked preponderance in the early afternoon and a preference for rain in the summer months over the continents and in the winter months on the oceanic coasts. But in the region between 40° and 60° of north latitude, which we shall come to regard as the frontier province of tropical air in the conflict between itself and polar air, with their very diverse characters, a diagram may suggest to us that the local incidence of rainfall is controlled by secondary causes and not directly by the sun's place in the heavens.

In a publication of the Meteorological Office issued in 1899 we find a series of tables of the average hourly values of the meteorological elements at four observatories, Aberdeen, Valentia, Falmouth and Kew, from which we can construct diagrams not only of the amount of rain that corresponds normally with each hour of each month; but also of the number of times that rain was recorded in the several hours. From them we may extract some interesting information.

The table overleaf for the allowance of rain for each hour of each month at Kew Observatory, Richmond, tells us that, on the average of the 25 years, the hour with the least rain has been 11 o'clock in the forenoon, meaning 10h 30 to 11h 30, with its annual contribution of 22 mm. The rainiest hours are marked 13 to 17 in the table, three of them with 30 mm per annum and the other two with 28. So year in and year out the most rainy hour gets less than half as much again as the least. And looking at the produce of the hours in the several months the record for dryness is won by 11 and 12 in March

Rainfall at Richmond (Kew Observatory) in millimetres for each hour of the day of each month of the year—averages for 1871–95

Hour	Jan.	Feb.	Mar.	Apr.	May	June	July	Aug.	Sept.	Oct.	Nov.	Dec.	Year
a.m. 1	1·6	2·0	1·3	1·7	1·6	1·7	2·3	2·2	3·4	3·8	2·5	2·1	26
2	1·6	2·2	1·6	1·6	1·6	1·8	2·7	2·6	2·6	3·2	2·7	2·0	26
3	2·0	1·8	1·5	1·6	2·0	1·5	2·5	1·7	2·9	3·4	2·5	2·2	26
4	2·1	1·6	1·5	1·9	1·7	1·5	2·2	1·4	3·3	2·7	2·3	2·2	24
5	2·0	1·8	1·7	2·0	2·3	1·8	2·3	2·3	3·0	2·9	2·5	2·4	27
6	2·1	2·5	2·2	2·1	2·1	1·9	2·0	1·7	2·1	3·5	2·5	2·2	26
7	2·2	1·6	2·1	1·9	2·2	1·9	2·6	2·5	2·0	2·9	2·4	1·9	26
8	2·5	1·9	1·5	2·0	2·1	1·8	1·7	2·2	1·8	3·3	2·3	2·0	25
9	2·5	1·9	1·5	2·0	1·5	1·7	1·8	2·2	2·0	3·2	2·0	2·3	25
10	2·4	2·0	1·7	2·3	2·0	2·2	2·7	1·9	2·0	3·4	2·2	1·6	26
11	1·8	1·7	*·7*	1·5	1·5	1·8	2·9	1·8	2·0	2·8	1·8	1·6	*22*
Noon 12	1·7	1·4	*·9*	1·5	2·1	2·5	3·8	2·5	1·9	3·9	1·8	1·9	26
13	1·8	1·9	1·4	2·1	1·8	1·5	**5·7**	2·7	2·2	4·5	2·3	1·9	**30**
14	1·9	1·9	1·3	2·0	2·3	1·8	4·2	3·5	1·8	3·0	2·4	1·7	28
15	2·6	1·6	1·5	2·0	2·8	2·4	4·5	3·3	1·5	3·3	2·9	2·2	**30**
16	2·1	2·1	1·2	2·3	2·5	2·4	4·0	2·8	1·7	3·2	2·5	2·2	28
17	2·2	1·4	1·2	2·3	2·8	2·6	2·9	3·6	2·6	3·0	3·5	1·8	**30**
18	2·2	1·7	1·8	2·1	1·9	2·3	3·5	2·6	2·0	2·6	2·5	2·1	27
19	2·0	1·5	1·2	1·5	1·2	2·9	2·5	2·8	1·9	3·4	2·7	2·1	26
20	1·5	1·4	1·1	1·6	1·2	2·4	2·7	2·0	2·2	3·4	2·6	2·0	24
21	1·7	1·4	1·4	1·5	1·4	3·4	2·2	3·1	2·5	2·7	2·0	1·9	25
22	1·7	1·4	1·5	1·7	1·1	3·1	2·9	2·5	1·8	2·2	2·6	1·7	24
23	1·9	1·1	1·8	1·8	1·6	2·1	2·7	2·0	2·4	2·6	3·3	1·8	25
Midt. 24	1·7	1·7	1·2	1·7	1·7	1·8	3·0	1·7	3·0	2·3	3·0	2·4	25
	47	40	*35*	44	45	51	70	57	54	**75**	60	48	627

each with an average of less than a millimetre for the month. The record for wetness goes to 13h (12.30 to 1.30 p.m.) in July with an average of nearly 6 mm for that hour during the month. All these must be more or less accidental; the July maximum is easily accounted for by the violence of thunder rains, which happen sometimes in July and are very disturbing; but the remarkable dryness in the March morning has no easy explanation—it might even mean some misadventure with the arithmetic. The nearest to those exceptional values are three hours of just over a millimetre between 18h and midnight in February, March and May, and six of 1·2 mm four between 16h and midnight in March and two within the same range 19h and 20h in May. The table is full of irregularities which arise from the special character of a station in the frontier province and on the western margin of the greatest land area of the world.

Rainfall at Richmond, number of hours that gave rain.
Averages for hours in the months of the 25 *years* 1871–95

Hour	Jan.	Feb.	Mar.	Apr.	May	June	July	Aug.	Sept.	Oct.	Nov.	Dec.	Year
1	3	3	3	3	2	2	2	3	3	4	4	3	35
2	4	3	3	2	2	2	2	3	3	4	4	3	35
3	4	3	3	3	2	2	2	2	4	5	4	3	37
4	4	3	4	3	3	2	3	3	4	5	4	4	42
5	4	3	3	3	3	3	3	3	4	5	4	4	42
6	4	3	4	3	3	2	3	3	4	5	4	4	42
7	5	3	4	3	2	2	3	3	3	5	4	4	41
8	4	4	3	3	2	2	2	2	3	5	4	4	38
9	4	4	3	3	2	2	2	2	2	4	4	3	35
10	5	4	4	3	3	3	3	3	3	5	5	5	46
11	3	2	2	2	2	2	3	2	2	3	3	2	28
12	3	2	2	2	3	2	3	2	2	4	3	3	31
13	3	3	3	3	3	2	3	3	2	4	3	3	35
14	3	3	2	3	3	2	3	3	2	4	3	3	34
15	4	3	2	2	3	3	3	3	2	4	4	3	36
16	4	3	3	3	3	3	4	3	2	5	3	3	39
17	3	3	3	3	3	3	3	3	3	4	4	3	38
18	3	3	3	3	3	3	3	3	3	3	4	3	37
19	3	3	3	2	2	3	3	3	2	4	4	3	35
20	3	3	3	2	2	3	2	3	3	3	4	3	34
21	3	3	3	2	2	3	3	3	2	3	4	3	34
22	3	3	3	3	2	3	2	3	2	3	4	4	35
23	3	3	3	2	2	2	2	2	3	3	4	3	32
24	3	3	3	3	3	2	3	3	3	4	4	3	37
	85	73	72	64	60	58	65	66	66	98	92	79	878

If we tabulate the *number of hours* which showed rainfall, the table of so-called rain-hours which results is more manageable and we have set it out for the reader's inspection. The frequencies for the hours range from 5 for the small hours of October mornings to 2 for 8h and 9h, May to September, and for 11h in eight months out of the twelve and for many hours in the night.

Looking at the vertical column of totals for the hours in the year, there is a maximum group for 4, 5, 6, 7 a.m. and a minimum for 11, so we may infer that for the neighbourhood of London the proverb "rain at 7, fine at 11" has some claim to respect, and the inference is borne out by the several months except May, June and July.

It is only fair to say that the corresponding table for the ten years 1903–12 with only 711 rain-hours for a year tells a somewhat different story.

Differences are not very prominent; but October has a very rainy character, which we may express by saying that whereas the chances of using an umbrella on the average for the whole year are one in ten (878 hours out of a total of 8766), in October the chance is one in eight and in June one in twelve. For February which is nicknamed "fill-dyke" the chance is one in nine.

Of course there are not many stations that give information for the several hours—it requires not only a self-recording rain-gauge but the patience to tabulate the record for the hours. Still, if there is any really well-marked character in the incidence of the salient features of the elements of weather, we can use less elaborate tabulation. Every two hours or three hours, or every four hours, the normal period for observations at sea, would show any marked character and indeed the four observations of the Daily Weather Report might give some useful indication.

Geographical lines. Contours and isograms

If we place lines of equal temperature, or equal allowance of sunshine, or any other element, *at successive times* at a particular station, in a category of their own as isopleths, we cannot avoid the invitation to provide also a name for the lines that connect points on a map or on a vertical section which have the same temperature, pressure or whatever it may be. They are indeed indispensable for the atmospheric engineer to represent his plans, elevations and sections, and are perhaps better known than any other form of meteorological diagram as isobars for lines of equal pressure on a map, isotherms for lines of equal temperature, isohyets for lines of equal rainfall, isonephs for lines of equal cloudiness and quite a multitude of others. Sir Francis Galton included them all in one class as "isograms", and as representing the geographical distribution of pressure, rainfall, temperature, etc. they are the foundation of meteorological atlases and of the weather-map. Isograms may be said to carry a larger cargo of potted weather-information than any other pictorial device. Generally speaking, in dealing with normal values, the separate elements are represented on separate maps; but on the synchronous charts of the Daily Weather Reports which we shall discuss in

Chapter v provision is made for the representation of all the elements observed, of the pressure by isobars, of the winds by arrows, the weather by symbols and the temperature by figures.

The adequate representation of mean values by a series of charts would mean a meteorological atlas. We must content ourselves for the present with less and we offer only, as examples, in fig. 51 charts of the normal distribution of pressure in four of the twelve months of the year, namely April, July, October, January adorned with colours to indicate the position of the normally rainy regions and the normally arid regions of the earth in those months.

The distribution of pressure is sufficiently indicated by toothed lines, isobars, with the corresponding pressure in each case marked in millibars. The areas of normally rainy regions and normally arid regions over the land are also sufficiently indicated by colouring the rainy regions green and the arid regions brown. The boundary lines of the rainy regions are lines of equal normal rainfall (isohyets) of 100 millimetres in the month—within the boundary rainfall is greater than 100 mm, and those of the arid regions by the corresponding lines for 25 millimetres—within these boundaries rainfall is less than 25 mm.

Over the sea in the absence of sufficient measures of rainfall we have used the normal amount of cloud instead, with the somewhat arbitrary assumption that a region with more than seven-tenths of the sky normally covered by cloud might count with the rainy regions, and those with less than five-tenths of the sky so covered would count with the arid regions. The reader may be interested to compare the results with the diagrams of the normal rainfall in many parts of the world which are displayed on pp. 174, 175.

It will be seen that the regions coloured brown do comprise the better known desert regions of the earth and what used to be called the horse latitudes, a region fatal in days gone by for the transport of horses over the sea. And in like manner the green colour covers the areas which are noted for abundance of vegetation.

The combination may serve to give a general idea of the conditions of life in the different quarters of the globe from the point of view of a natural sufficiency of water.

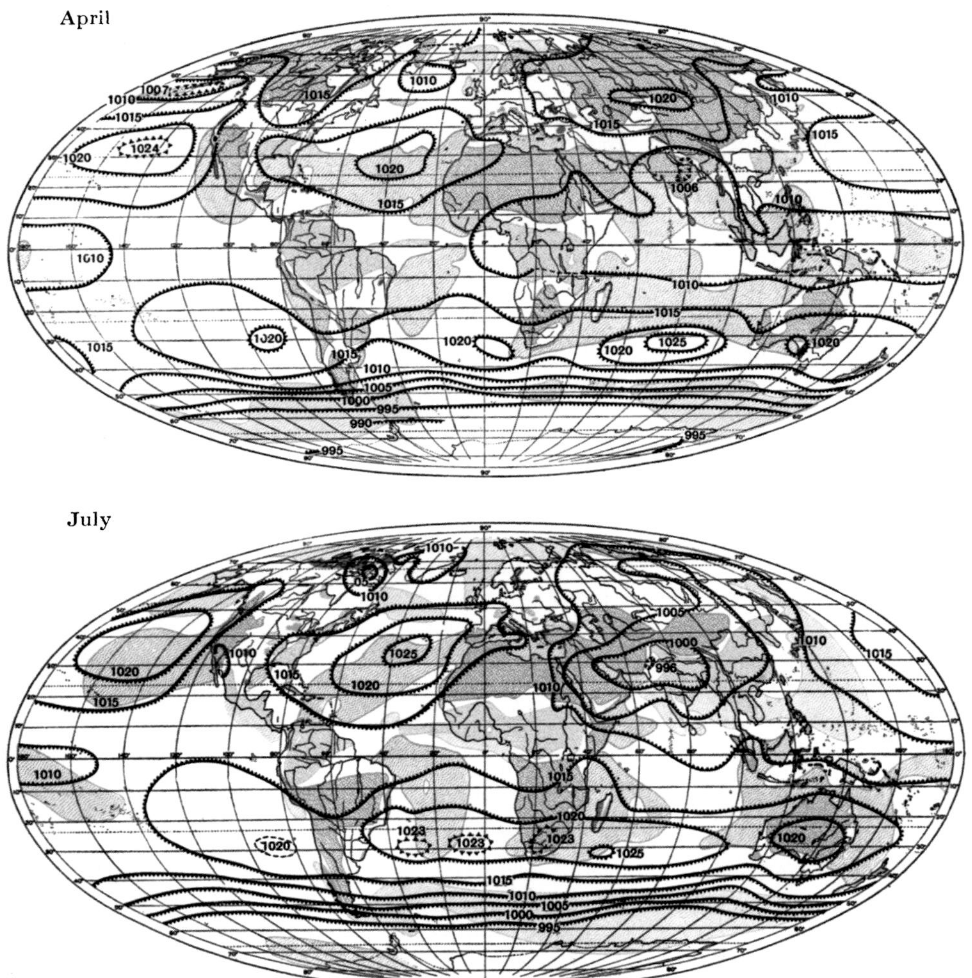

Fig. 51. The scheme of colour for the maps is as follows:

Double green stipple, regions of rainfall over the land normally greater than 100 mm, about 4 inches, in the month under consideration.

Single green stipple, regions of cloud over the sea normally covering more than seven-tenths of the sky. These are suggested as rainy regions.

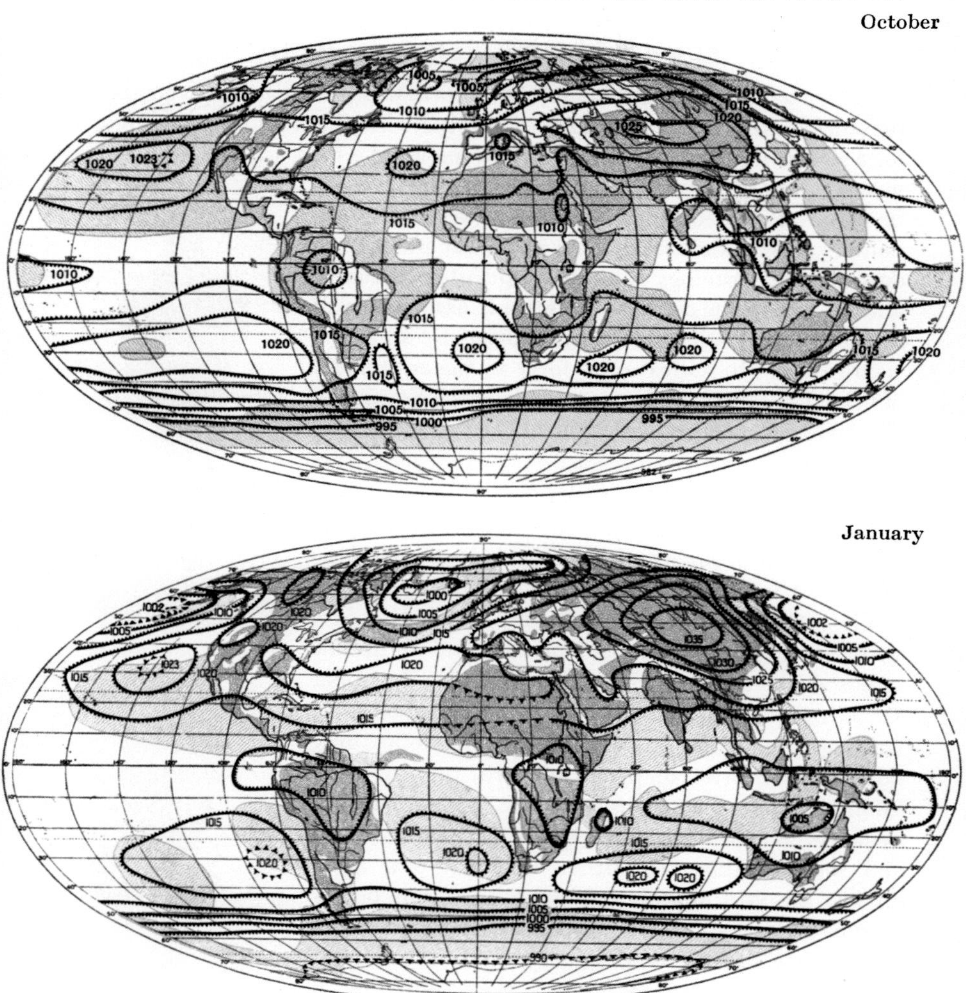

Fig. 51.

Double brown stipple, regions of rainfall over the land less than 25 mm, about 1 inch.
Single brown stipple, regions of cloud over the sea covering less than five-tenths of the sky.
These are suggested as arid regions.
The areas left white, which include the greater part of the British Isles, are regions of moderate rainfall.

THE REPRESENTATION OF WIND

So far we have considered mainly the information that can be given about pressure, temperature, sunshine or rainfall which require only one figure for a single observation. Wind requires more, one for velocity and in effect three for direction because, to be complete, direction has to allow for three components, North and South, East and West and up and down. In meteorology the last is not often available and for a map the direction of wind is assumed to be horizontal and adequately represented by an arrow with feathers to indicate the force as the equivalent of velocity. Horizontal components North-South and West-East are sometimes used instead.

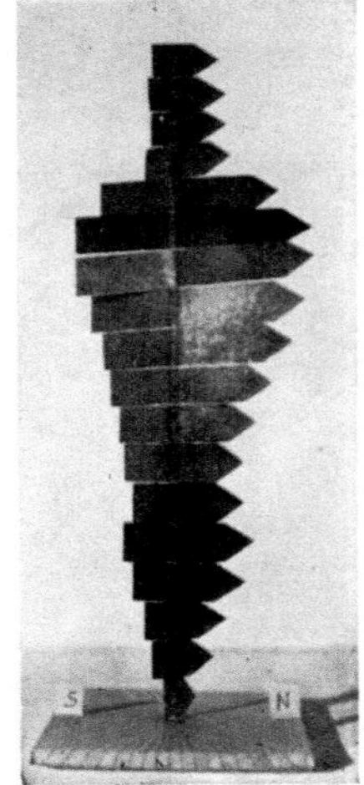

Fig. 52 *a*. 1 Oct. 1908.

Winds in the upper air

Perhaps as the wind is the one element that enjoys freedom of movement we had better begin with the winds of the free air.

The simplest of all pictorial representation is a scale-model of the subject of inquiry, and for us this is most easily realised in the case of winds in the upper air, which are identified by observations of pilot balloons. Neglecting any vertical motion, models may be made by a set of pieces of card, one for each kilometre with length proportional to the observed velocity, pointed to show the way of the wind, and set one above the other on a hat pin. Such models are very instructive when they can be inspected; they are not adequately expressed by photographs, because the camera only takes one aspect. Still, even with that restriction, they enable us to indicate some useful information, so we give four (fig. 52) which are taken from observations by C. J. P. Cave. The aspects chosen are (*a*) from East

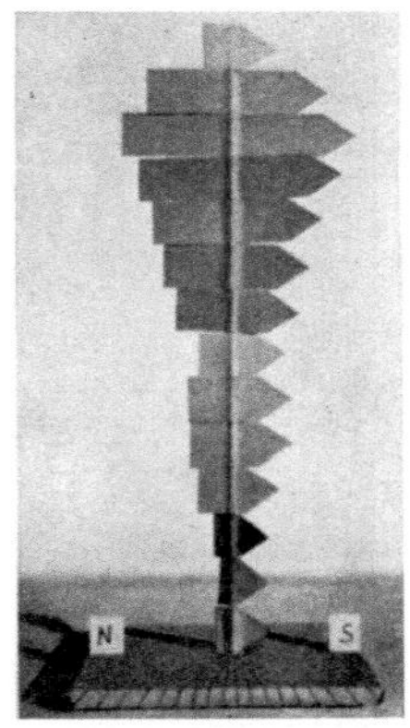

Fig. 52 *b*. 29 July 1908.

Fig. 52 *c*. 6 Nov. 1908.

Fig. 52 *d*. 1 Sept. 1907.

showing the South to North component of the motion of the air, or (*b*) from West showing the North to South component, or (*c*) and (*d*) from South showing a component from the East when the point of the card is on the left or a component from the West when the point is on the right.

With this understanding fig. 52 *a* shows a gradual increase of southerly wind from very little velocity at the ground to about ten times as much at the twelfth kilometre and then a rapid lapse to something not very different from the surface-layer; and fig. 52 *b* shows a very similar range in a northerly wind between the same levels.

On the other hand fig. 52 *c* shows a moderate wind from the east in the first kilometre dwindling to calm in the fourth and being replaced in the upper layers by wind from the western side which increases regularly from the fifth to the tenth kilometre. And fig. 52 *d* shows a very light wind from something west of north changing to a wind from the west increasing rapidly in the four kilometres through which the observations extended.

The elegant simplicity of representation by an arrow suggests the idea of expressing the normal or prevalent association of the winds on a map of the world using thickness of the line to suggest differences of force and the length of the line to suggest steadiness. On that plan maps of the prevalent winds of the globe for January-February and July-August are shown in fig. 53. They are based on maps of a different kind prepared by Wladimir Köppen, now of Graz, one of our veteran meteorologists.

It should be noticed that the winds are given only for the oceans, corresponding maps for the continents can hardly be hoped for because the winds are so much affected by inequalities in the surface. For them, in the figures referred to, we have indicated the sea-level isobars and would ask the reader's attention to the agreement between the line of the isobars and the line of flow of the winds at the coasts.

Jan.–Feb.

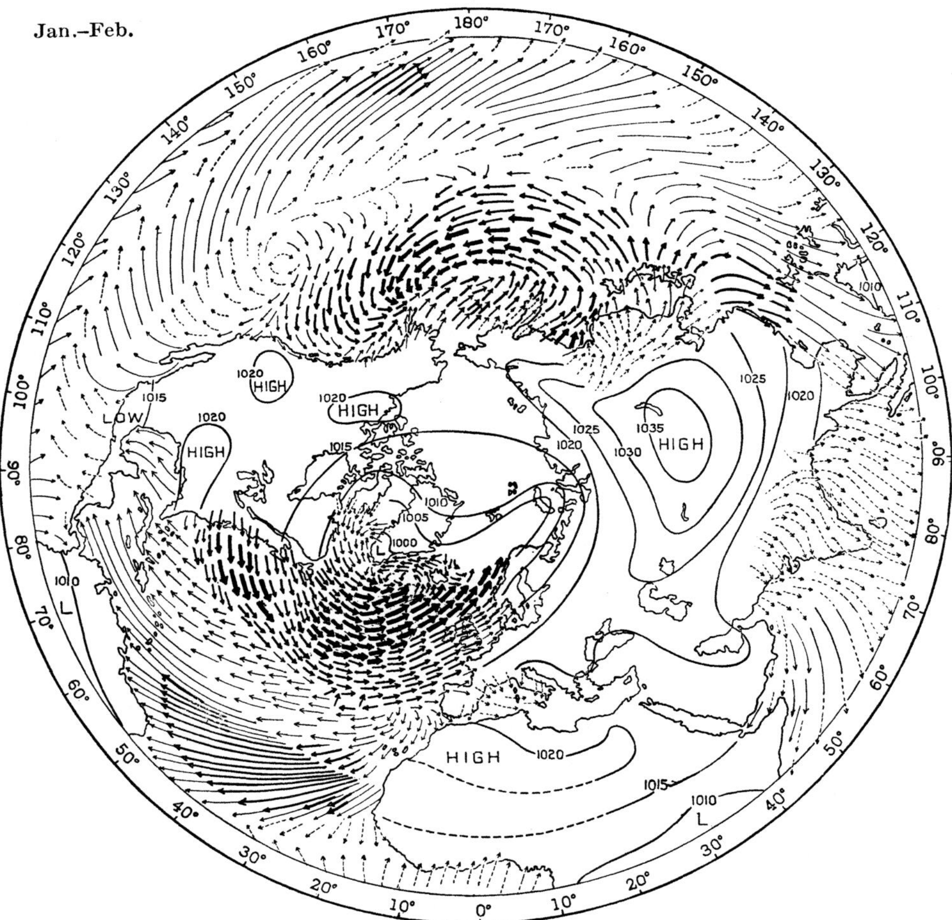

Fig. 53 *a*. Prevailing winds over the oceans of the northern hemisphere in January-February (after W. Köppen). The arrows are drawn in the direction of the prevalent wind. They show by their length the steadiness of the wind, and by their thickness its force on the Beaufort scale according to the following convention:

····→ under $3\frac{1}{2}$, ⟶ $3\frac{1}{2}$ to $4\frac{3}{4}$, ⟶ $4\frac{3}{4}$ to 6, ⟶ over 6.

July–Aug.

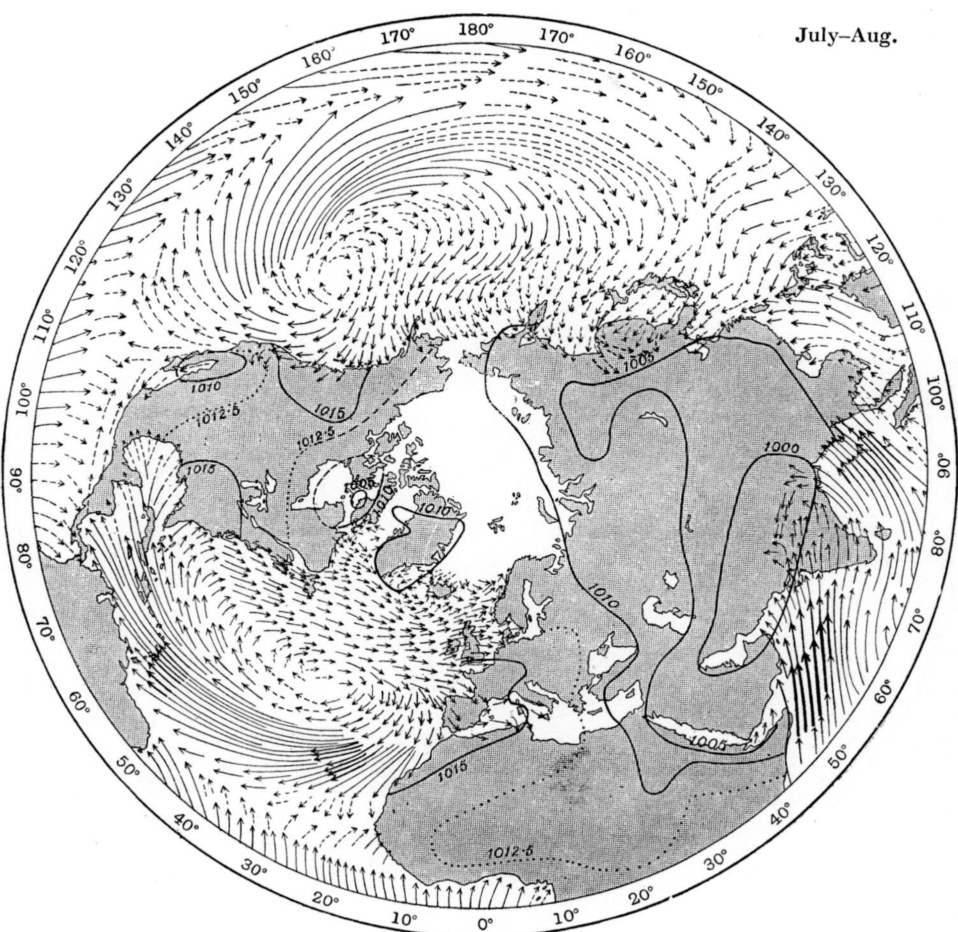

Fig. 53 *b*. Prevailing winds over the oceans of the northern hemisphere in July–August (after W. Köppen). Wind-arrows as in 53 *a*. Land-areas are shaded.

Please look for the north-east trade-wind, the monsoon, the high pressures of the Atlantic and Pacific and for a low near Iceland. (See pp. 156–7.)

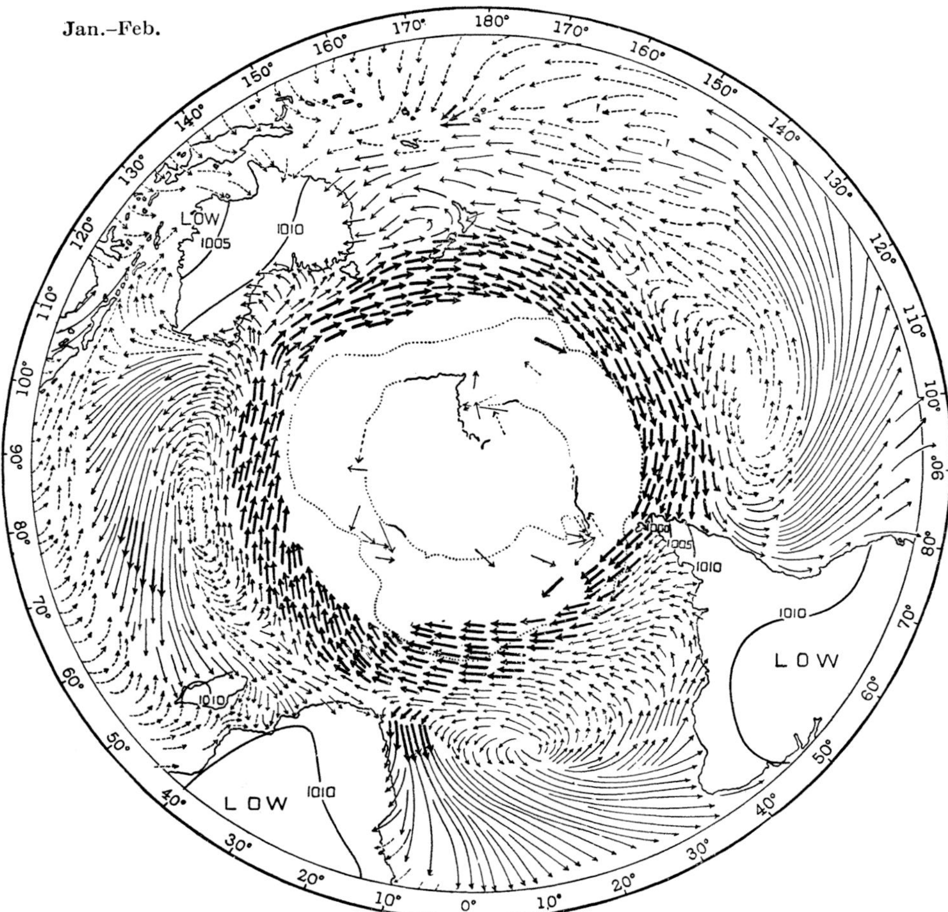

Fig. 53 c. Prevailing winds over the oceans of the southern hemisphere in January–February (after W. Köppen). The arrows are drawn in the direction of the prevalent wind. They show by their length the steadiness of the wind, and by their thickness its force on the Beaufort scale according to the following convention:

---→ under $3\frac{1}{2}$, → $3\frac{1}{2}$ to $4\frac{3}{4}$, → $4\frac{3}{4}$ to 6, → over 6.

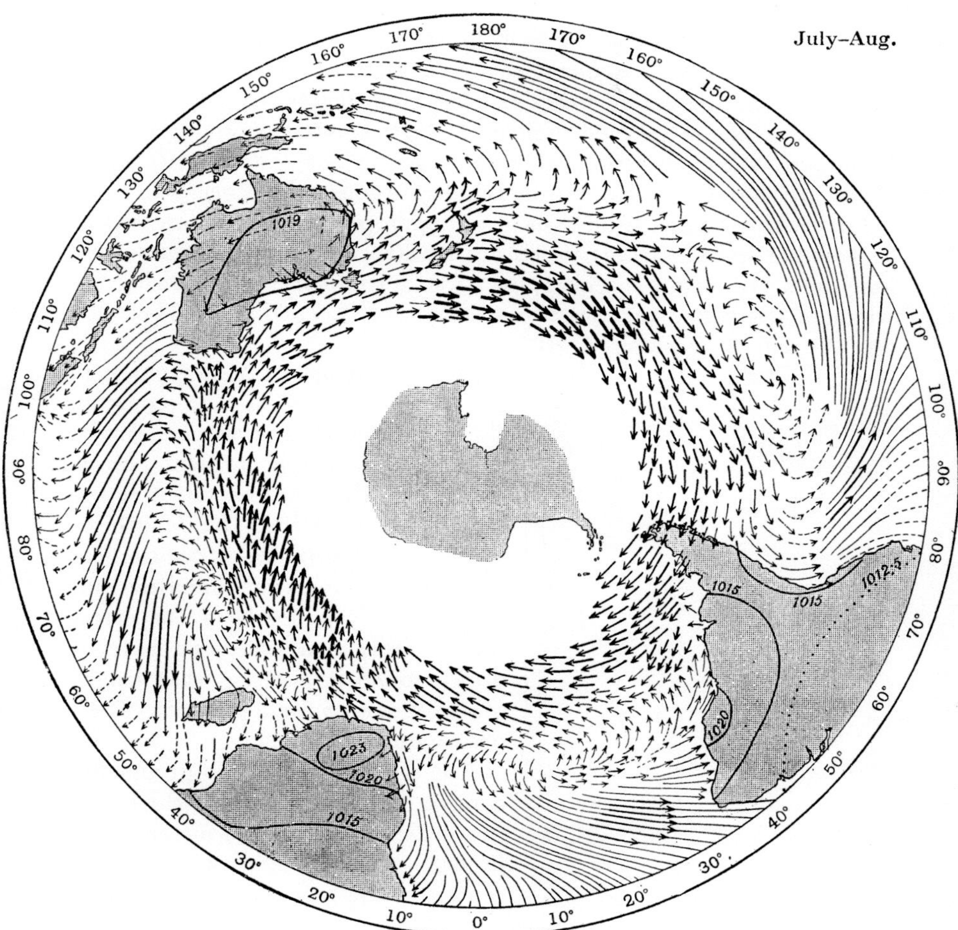

Fig. 53 *d*. Prevailing winds over the oceans of the southern hemisphere in July–August (after W. Köppen).

Very like 53 *c*. Please look for the south-east trade-wind and its modification on the north-west coast of Australia, the high pressures of the three oceans by the suggestion of counter-clockwise circulation, also the Roaring Forties.

Winds at a single station. Wind-roses and stars

Such maps are very eloquent of the characteristic features of the circulation of the air over the world's surface; but they tell us only about the prevalent winds at selected periods of the year without giving us anything more than suggestions about the cross-winds and opposite winds which are characteristic of all the temperate zones of the oceans.

For these we have wind-roses quite appropriate for land-stations as well as for ocean-squares which are marked by steps of latitude and longitude. We give a specimen, fig. 54, of a wind-rose summarising the winds of the whole year for the sea west of the Scilly Isles between lat. 48° and 52° N, long. 5°–10° W.

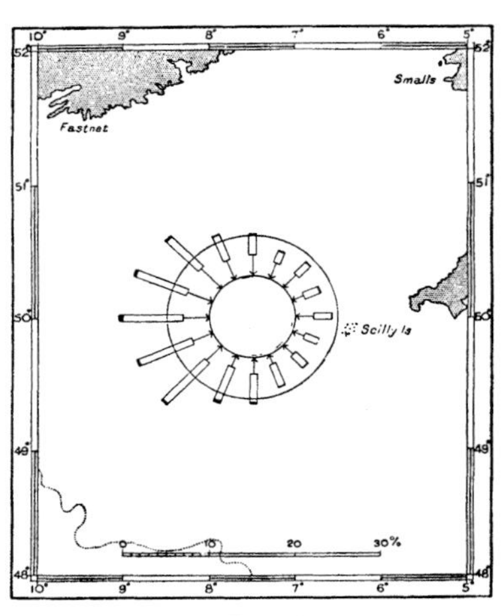

Fig. 54.

The experience of wind from each of sixteen points of the compass (N, S, E and W and three between each pair of those cardinal points, NNE, NE, ENE and so on) is represented by an arrow the length of which shows the number of occurrences of that wind for every hundred observations. The points of the arrow toe a circle and another circle is drawn round it to show what length of arrow counts for 5 per cent of the observations.

The observations are in groups according to the numbers of the Beaufort scale of wind-force, forces 1 to 3 in one group of light winds represented in the rose by a line leading to the point of the arrow, forces 3 to 7 in the group of moderate or strong winds represented by an open double line. At the ends of these are small sections blackened; they represent the number of gales, force 8 or more, in each hundred observations. In the diagram not more than one per cent of the whole number of observations for any wind-direction are gales. The W, WNW and NW are most distinguished in that respect.

The general idea of a wind-rose is to tell us how many observations out of every hundred recorded for each direction belong to the separate classes light winds, moderate or strong winds and gales. The observations not accounted for in this way are calms. For fig. 54 the winds of the whole year have been grouped together within forty-eight compartments, allowing three for each of the sixteen directions in each month; but we may use the same process for the observations made in each of the twelve months and so sort out the whole information into 576 compartments with twelve additional compartments if we wish to recognise the calms of each month.

Thus we might express the year's experience in twelve wind-roses, one for each month on the same plan as fig. 54 for the whole year, or all together in a composite rose or star as in fig. 55.

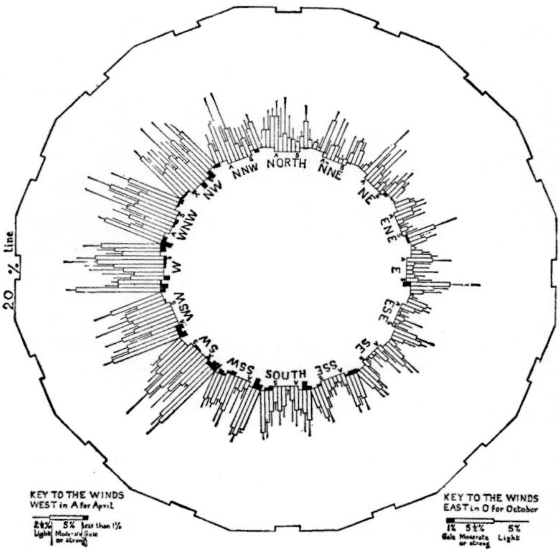

Fig. 55.

Fig. 55 shows a normal year's contribution to all these separate 576 compartments but arranged in a star, rather differently from the plan of the rose in fig. 54, inasmuch as the double lines representing moderate or strong winds for the successive months are arranged side by side to toe an indented line which faces the direction from which the winds came; the single lines for light winds are outside the moderates instead of within and the blackened columns representing gales project into the interior of the figure furnished by the indented lines facing the sixteen wind-directions. The indentation of the base line allows us to distinguish the information for the summer half, April to September, for the northern hemisphere, from the two winter quarters, January to March on the left for a spectator looking from the centre and October to December on the right. Small letters A and S, hardly visible in the reproduction without a glass, mark the ends of the salient.

A sufficient number of stars of the kind represented in fig. 55 would enable us to set out the regime of winds for the world. But the practice is to give the information in a dozen separate charts of wind-roses one for each month, leaving the reader to express any seasonal variation as best he can.

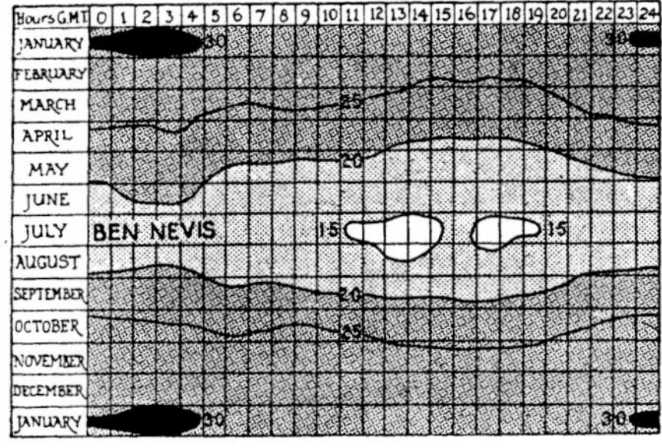

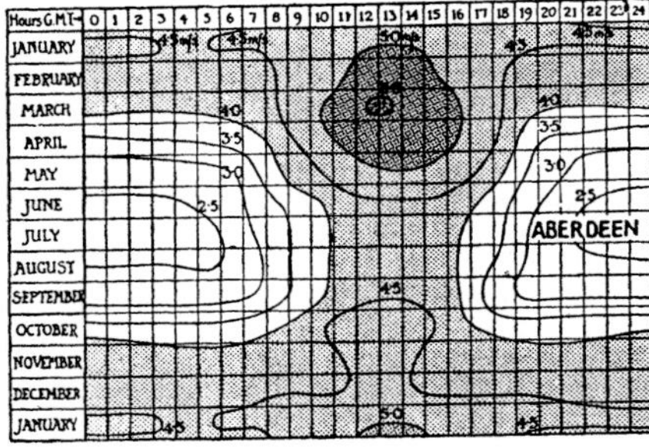

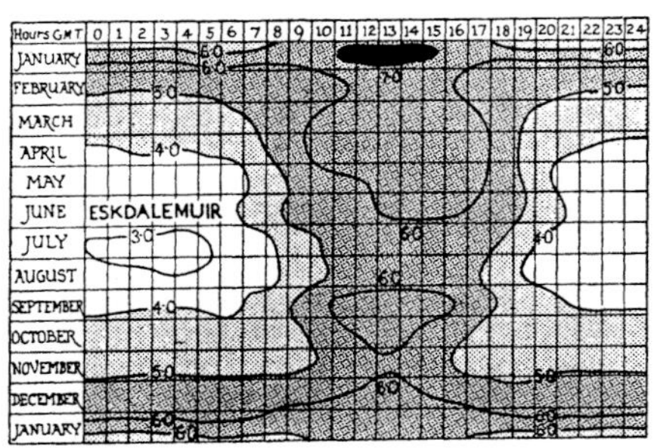

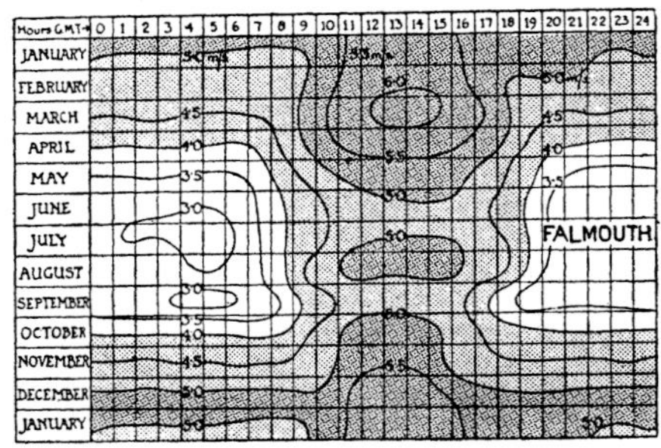

Fig. 56. The diurnal and seasonal variation of wind-velocity on Ben Nevis, and at Aberdeen, at Eskdalemuir in Dumfriesshire and at Falmouth.

The figures for Ben Nevis are on a special scale. 1·5 is approximately 9 mi/hr (4 m/sec), 2·0 12 mi/hr, 2·5 16 mi/hr, 3·0 21 mi/hr. For the other diagrams the lines are drawn according to the number of metres per second marked on the diagram. 1 m/sec is just under 2¼ mi/hr.

Lines of equal wind-velocity

Wind-roses are not designed to give us information about the variation in the wind within the 24 hours; and indeed at sea, for which wind-roses are chiefly required, there is little if any variation to record; but on land it is different—there the wind generally falls off at night and it is a disturbing experience for the wind to "go bellowing through the night till dawn". To exhibit this isopleth diagrams are the best, although they give results only for velocity: there is nothing about direction. Here are the diagrams for Ben Nevis, Aberdeen, Eskdalemuir and Falmouth (fig. 56). Ben Nevis is very orderly, ranging between a wind marked 1·5 (of a special scale) for the afternoon of July to a group of squares with force 3 on the same scale between 11 p.m. and 5 a.m. in January. Light winds by day, strong by night. Something like it is shown in fig. 57 for the top of the Eiffel Tower in June with its wind less than 6 m/sec at 9 a.m. and greater than 9 m/sec at midnight; but the others are quite different. At the foot of the Eiffel Tower there is a black maximum in the early afternoon in February and March, a pale shading only with an isolated calm period in September. At Eskdalemuir the range is orderly, from 3 m/sec in the July night to 6 m/sec in the early afternoon, and reaches 7 in January days. Aberdeen has relatively calm nights and a vigorous centre of 5·5 for March afternoons, and Falmouth equally calm nights, or nearly so, and a similar centre of 6 m/sec in March afternoons.

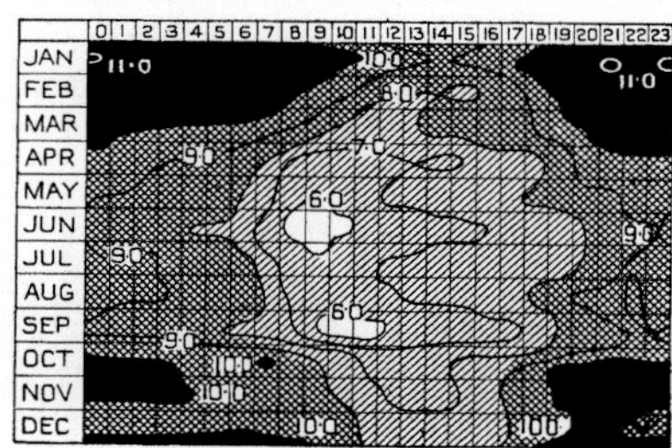

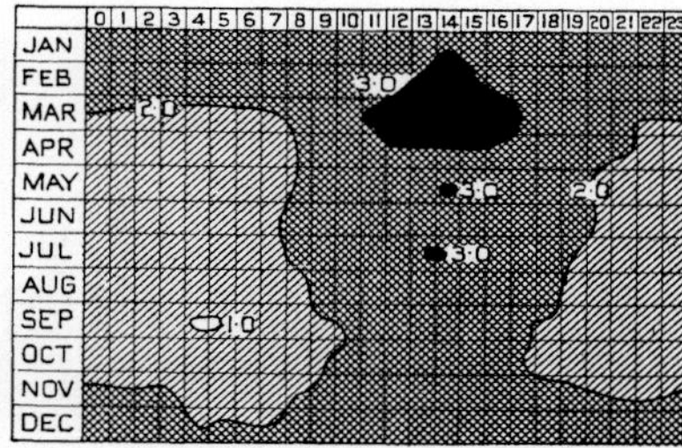

Fig. 57. The Eiffel Tower. Diurnal and seasonal variation of wind-velocity at the top (about 1000 ft) on the left, and on the right at the Bureau Central near its foot.

The figures indicate velocities in metres per second according to the following shades: *Eiffel Tower*: less than 6·0 m/sec white, greater than 10·0 m/sec black. *Bureau Central*: less than 1·0 m/sec white, greater than 3·0 m/sec black.

The most notable feature of the surface-winds is the relatively calm summer nights and disturbed summer days.

It expresses the fact to which attention was directed long ago by Espy in the United States and by Köppen in Germany that the cooling of the ground at night in consequence of the loss of heat by radiation to the clear sky generally introduces what may be called an artificial reduction of the wind-velocity which is made up by a fresh supply of heat from the sun when it has risen. "The moonlight sleeps, the sun awakes to turmoil." Hence we find that for most purposes afternoon is a better time for wind-observation than early morning.

STEP-DIAGRAMS

In the method of representation by graphs we deal with lines which express the continuous variation such as is actually recorded on a barograph, thermograph or hygrograph. For isopleths we draw lines of equality which wander between the normals marked for the hours of the several months as though we were assured of the position in which the particular value would be found which is appropriate to the line.

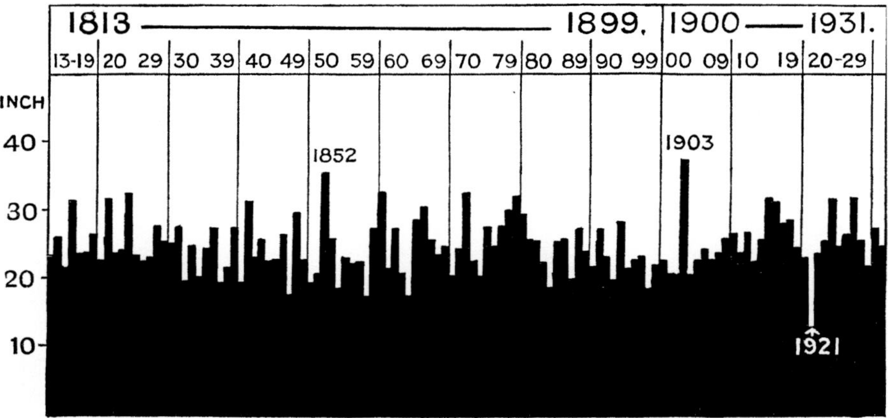

Fig. 58. The rainfall of London for each year from 1813 to 1931. The height of each black column, read on the scale on the left-hand side, represents the rainfall for a year. The column on the extreme left shows the rainfall for 1813, that on the extreme right the rainfall for 1931. Vertical lines are drawn across the diagram at intervals of ten years.

Here for summarising the results of observations we may take the liberty of commending another form of diagram the principle of which is already exemplified in our representation of the seasonal variation of the winds in so far as it presents a rugged discontinuity instead of a mild continuity. We give this form the name of step-diagram because it will serve the purpose of representing the steps by which changes take place from hour to hour, from day to day, from week to week, from month to month, or from year to year.

London rainfall in 119 *years*

If, for example, we have a series of values of rainfall year by year, such as that for the rainfall of London in fig. 58, we could at appropriate intervals mark on paper *single points* representing each year's contribution, and connect the points by a continuous line with suitable twists and turns; but the connecting lines drawn between the points would really be drawn on the imagination. An intermediate value between two yearly aggregates of rainfall has no meaning, whereas if we set up a column with its height proportional to the yearly rainfall, and leave it, it tells the truth, the whole truth and nothing but the truth, about the total rainfall of the year. Continuous curves are often drawn to form a graph by connecting together successive normals, as those for successive positions along a line of latitude in fig. 47, or those for successive weeks of the year in fig. 48, and there is no serious objection to using imagination for a sketch-book instead of information.

Let us look at some typical step-diagrams.

Sunshine and solar radiation week by week

Fig. 59 gives us, in its lower panel, the money value at an estimated price per Board of Trade unit of daily averages of the energy of solar radiation on a square metre of land at Rothamsted for successive weeks of the year 1924, with a line to indicate the equivalent of 50 per cent of possible.

In the upper panel the weekly duration of sunshine for the same year is represented by columns for the average number of hours per day; multiplied by seven they would tell us the actual duration for the week.

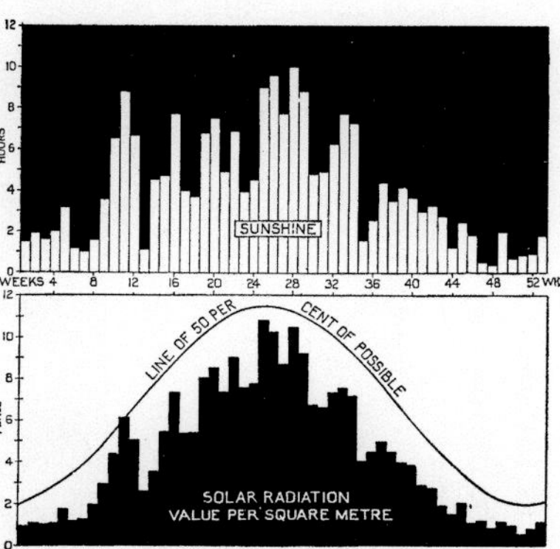

Fig. 59. Sunshine and solar radiation week by week at Rothamsted, Herts, in 1924.

Overleaf we give some information about the frequency of thunderstorms, similarly arranged.

Thunderstorms

As an example based on months let us give the diagram (fig. 60) which shows the normal distribution of thunderstorms throughout the year at 54 places in different parts of the world, and the diurnal experiences of incidence at two of them.

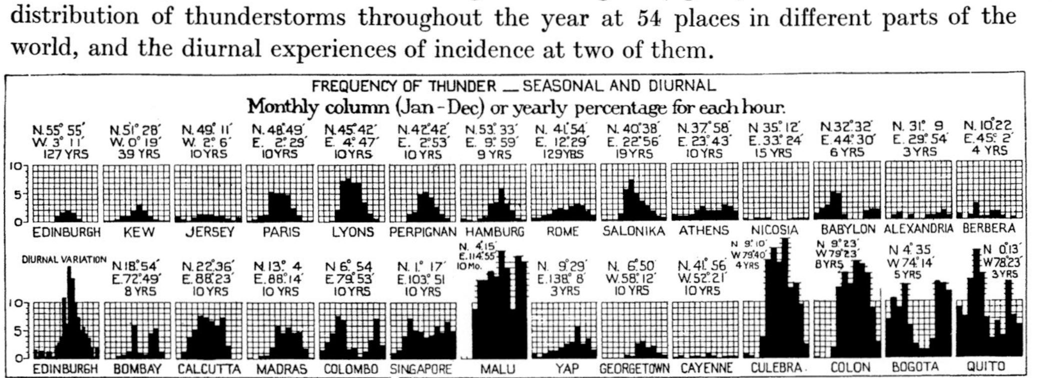

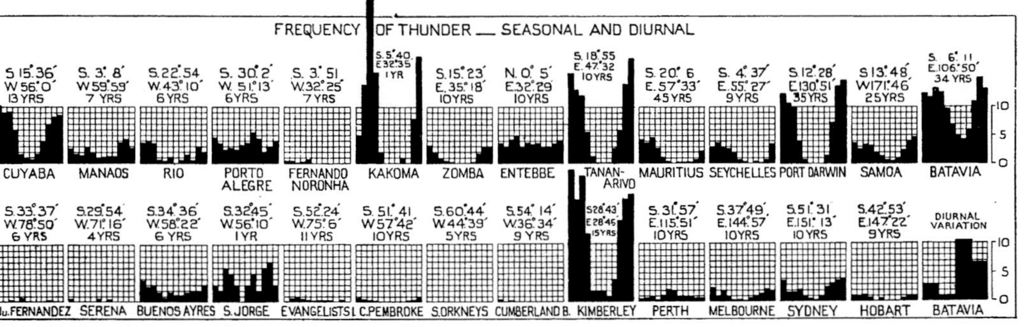

Fig. 60. Step-diagrams of the normal frequency of days of thunderstorm in each of the twelve months of the year at 27 stations in each hemisphere and the distribution of the storms within the day at Edinburgh in the northern and Batavia in the southern.

In the 54 diagrams of seasonal variation a column is drawn for each month and the scale at the side shows the normal number of days of thunder in the month. In the two diagrams of diurnal variation that at Edinburgh has 24 columns, one for each hour, that for Batavia four columns representing observations at 6-hourly intervals; in both cases the scale gives the hour's percentage of the total number of days of thunder.

TIME AND GEOGRAPHY ON THE STAGE TOGETHER

We shall illustrate the use of the step-diagram of mean values by employing it to represent some of the characteristic features of the distribution of rainfall, pressure and temperature throughout the year and over the globe.

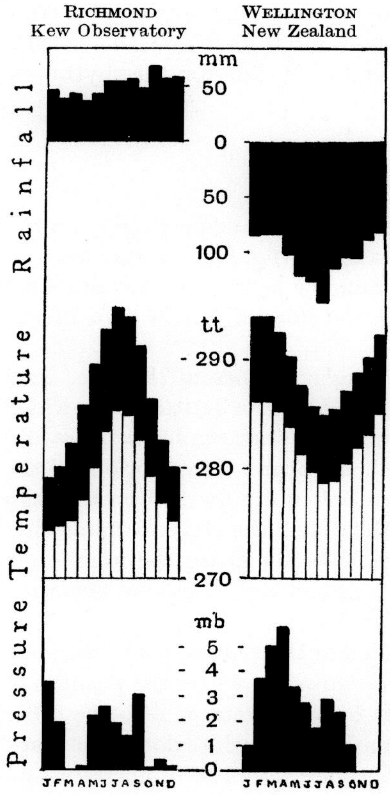

Fig. 61. Step-diagrams of rainfall, pressure and temperature at Richmond (Kew Observatory) on the left and at Wellington (New Zealand) on the right.

Each of the six diagrams is made up of twelve columns one for each month from January on the left to December on the right.

In the upper panel the length of each column represents the average monthly rainfall; the columns for Kew are drawn upwards, those for Wellington downwards from the same zero line, for a reason that will appear later.

In the panel marked temperature the top of each white column represents the average daily minimum temperature, the top of each black column the average daily maximum temperature in each month representing together the warmth of the day and the chill of the night. The length of the black column shows the average difference between the highest and lowest temperatures of the twenty-four hours.

In the lower panel the horizontal line marked 0 represents the pressure in the month which shows the lowest value of the twelve—March at Richmond, and November and December at Wellington. The heights of the columns show the excess of the normal pressure in the other months above the zero value.

In fig. 61 are six step-diagrams: representing in the upper row, the normal rainfall for each of the twelve months, below that the normal maximum and minimum temperature and at the bottom the normal pressure. The diagrams on the left-hand side are for Kew Observatory, Richmond, those on the right for Wellington, New Zealand, which is very near the antipodes of Kew.

We can recognise at once a rhythmical sequence in the maximum and minimum temperature at both positions; and the rainfall at Wellington, indicated by an inverted diagram, shows also a rhythmic tendency. But there is only a very faint rhythm in the rainfall for Kew; the normal pressure at Kew is not at all rhythmic and at Wellington the rhythm of pressure is a complicated one.

Rainfall

Diagrams of this kind can be used to bring time and geography on the stage of the world together. Our first example (fig. 62) is a new view of the distribution of rainfall in which a number of stations are provided each with a step-diagram representing the normal fall of rain in the months of the year.

The diagrams for the stations in successive zones of latitude, each covering ten degrees, are arranged to stand on the latitude-line which forms the southern boundary of the zone in the northern hemisphere and the northern boundary in the opposite hemisphere. The positions of the stations from which the observations are taken are marked on the map by round dots with numbers against them and the step-diagrams that belong thereto carry the same numbers. To save space the numbering begins afresh for every zone at the western end and carries on along the zone to the eastern end.

It is not always easy to fit the diagrams to their appropriate stations; the map on the small scale is rather over-crowded; where the position of a station is covered by the black of its own diagram or of some other station's, the position of the station is marked by a white dot in the black area and even that is not sufficient; there are numbers of diagrams, including, indeed, the one for Cherrapunji, the rainiest station of them all, for which room cannot be found within the limits of the map; they are accordingly set below the oval of the boundary with a reference number to the appropriate zone, and the numbered position within it. The numbers assigned to the zones are on the left-hand side of the oval and on either side of the diagrams at the foot of the picture. In that way the step-diagram belonging to a station with a particular number in a particular zone can ultimately (with some difficulty) be traced. The difficulty may be forgiven in consideration of the enormous amount of information which is contained in such small space and which represents an impressive co-operative effort involving millions of personal observations.

The scale of the columns indicating the month's rainfall is one-thirtieth of what may be called natural size, that is to say, a millimetre of black column represents 30 mm of rainfall, a third of an inch represents 10 inches; and the year's rainfall at any station is represented by the area of the black of the diagram—a square millimetre is 90 mm of rain.

With this explanation let us notice some of the salient features of the chart. First there are some places about the equator and the coast of southern Chile where it may be said to rain heavily all the year round. Examples of tolerably uniform rain throughout the year can be found also on the east coasts of the United States and Canada and the western coasts of the British Isles, the shores of the Northern Atlantic; but looking at the matter in greater detail a hump in the middle of the diagram shows rainfall as a summer visitation over the great Eurasian continent and over the North American continent. In the southern hemisphere the corresponding statement may be made; rainfall is prominent at the beginning and end of the year, but in that hemisphere those are also the summer months. So that we may conclude that over great land-areas the summer is the rainy season. The opposite is the case with coast-stations and island-stations as is evident by the behaviour of the diagrams at the beginning and end of the year in the islands and eastern shores of the Atlantic and on the western coasts of North America and in the middle months of the year on the western coasts of South America, which are the winter months of that region. The Mediterranean too shows rain in the winter.

Notice, also, patches of coast devoid of any diagrams, western South America down to the tropic, western South Africa and western Australia over the corresponding region. Those with Sahara and Arabia and parts of North America and Central Asia in corresponding latitudes of the northern hemisphere are the rainless regions. Central Australia shows another for the southern hemisphere.

In their turn pressure and temperature over the globe will be similarly represented. On pp. 295–300 an index of the stations gives the positions from which the information has been derived.

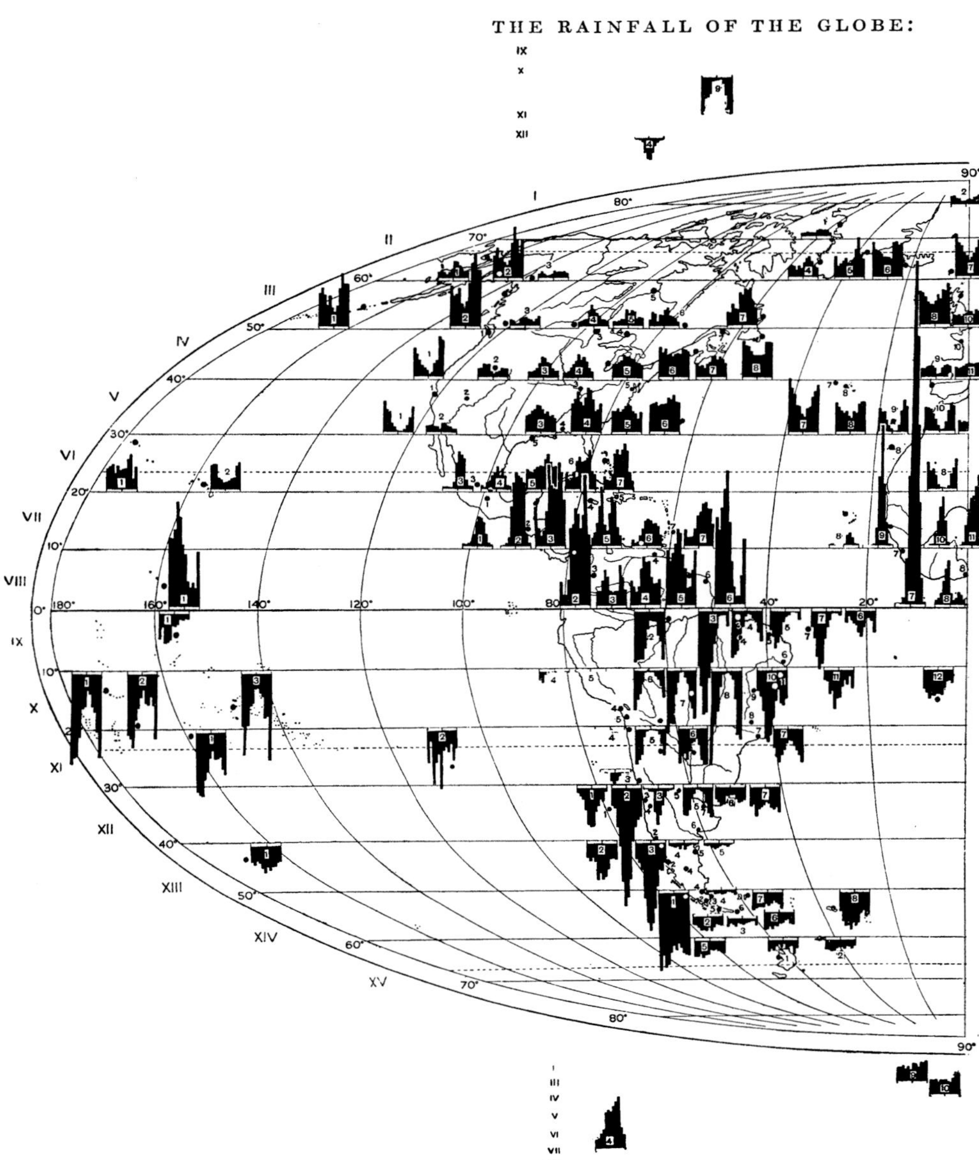

Fig. 62. Diagrams of the sequence of normal rainfall for the twelve months of the year at 280 stations. The columns for the successive months January to December run from left to right in either hemisphere. They are drawn upward for stations north of the equator and downward for stations south. The scale is 1 mm for 30 mm of rainfall. Two specimens of the diagrams, on a larger scale, are given in fig. 61.

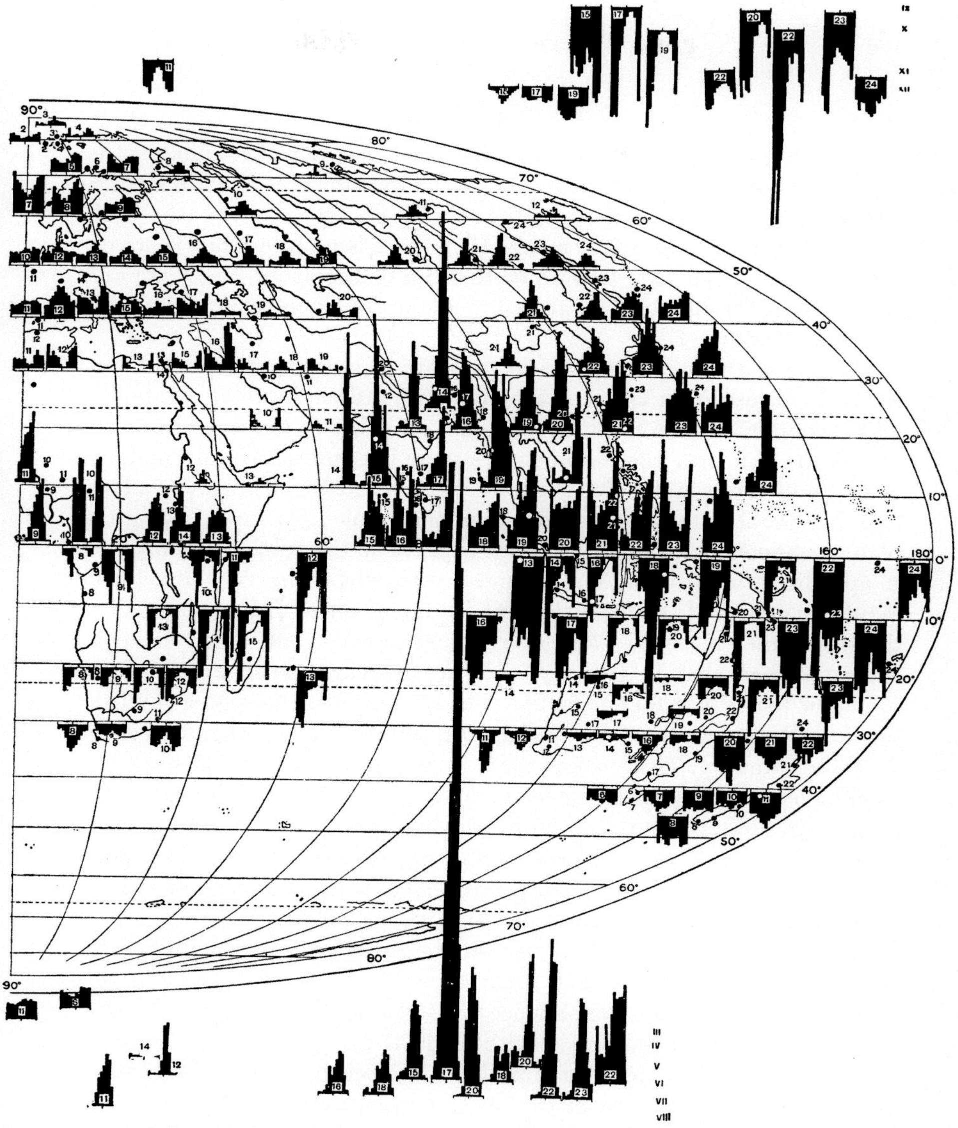

The positions of the stations are marked by black dots (in some cases by white dots). Those in each zone of ten degrees of latitude are numbered consecutively from west to east and the same number is marked on the corresponding diagram. The diagrams are ranged along the base line of their zone. Those for which there is no room in the line are set in the lower or upper margin, those of the northern hemisphere below their place, those of the southern hemisphere above. The zones to which the outer diagrams belong are indicated right and left.

Pressure over the globe

We have treated pressure (fig. 63) in a manner somewhat similar to that of rainfall but with this difference—the columns for rainfall indicated the whole depth of water reached by the rain which fell in the month. If the month shows the zero line it means that there is no rain. But there can never be any condition of no pressure or no temperature, there can only be differences from month to month made up by aggregating differences from day to day.

The use of the step-diagrams of pressure at stations distributed over the world arose from the fact that if we regard the normal pressure of 29·5 inches or 750 mm near sea-level as equivalent to the weight of a ton per square foot or a kilogramme per square centimetre, we can make an estimate of the total weight of air over the whole hemisphere by adding together the loads on separate parts of the whole area. Employing this plan to estimate the weight of the atmosphere on the northern hemisphere it appeared that the weight changed with surprising regularity from month to month of the normal year. In the month of July which follows the mid-summer solstice there were 10,000,000,000,000 tons less air on the hemisphere than in January following the winter solstice; so we have somehow to account for the fact that those ten billion tons go away from the north between January and July and come back home again between July and January. That means that there is an enormous transfer of air and the monthly values of pressure ought to tell us what happens to the air-load in different parts of the world in the course of a normal year. So we may present step-diagrams of pressure after the fashion of the step-diagrams of rainfall, only remembering that it is the variation of load we are now representing; the total need not be shown.

The lengths of the columns are on the same scale as the movement of a mercury barometer, inch for inch.

Here, contrary to our experience with the rainfall, we find the continents, especially Asia, loaded with air in the winter and the adjacent oceans north of the tropic have their load much lightened. So that we must understand that one of the operations of nature is to remove air from the oceans to the continents for the winter and replace it for the summer; but we have still to remember the ten billion tons which the northern hemisphere loses by export to the southern; and looking at the diagrams for the southern hemisphere we may be surprised to see that there is no such law of exchange between the land and the water as we find in the northern. Up to the 40th parallel at least the land and water seem to be quite of one mind about the hoarding of air.

It is a suggestive result but we cannot pursue it off-hand, because our diagrams for latitudes beyond 40° tend to become irregular and display a lack of sufficient data to justify deliberate statements.

In the map the changes of pressure from month to month are represented by the differences in length of columns, one for each month, drawn from a base line indicating the level for the month when the pressure is lowest. As with the rainfall-map the whole area is divided into zones of ten degrees of latitude and the stations within each separate zone are numbered from west to east. The diagram for each station that can be accommodated within the confines of the map is centred in the station's ten-degree square with the line of mean pressure set along the middle line of latitude of the zone. And again diagrams that cannot be accommodated on the map are set in the margins. Those for the northern hemisphere in the upper margin and those for the southern hemisphere in the lower margin; and again the mean line as well as the diagram is numbered with the number of the zone to which the station belongs. The longitude of each station is indicated in the index on pp. 296–300.

On this map three-quarters of a millimetre will indicate a millibar of pressure, or approximately the weight of a gramme per square centimetre, a thousandth of a ton per square foot or a ton per square dekametre, and in the case of the atmosphere it is the weight of the local column of air which causes the pressure.

So we may regard the diagrams as indications of the seasonal export and import of air between the different localities and our pressure-diagrams as statistics of the seasonal traffic of air between the locality of the stations and the rest of the world. The traffic is very different in different localities and our diagrams suggest the difference and invite us to think of the causes.

THE BAROMETER AS TRAFFIC INDICATOR

One millibar is represented by three-quarters of a millimetre, or by three hundredths of an inch.

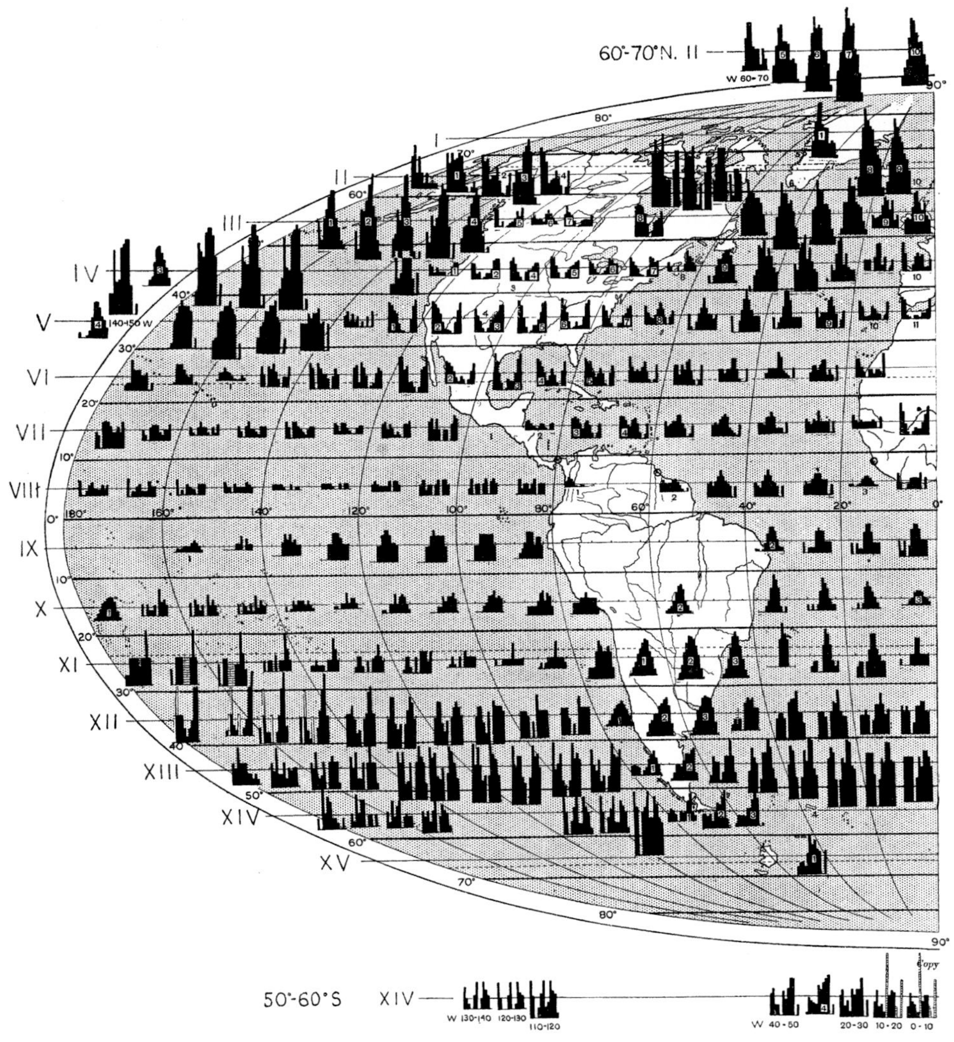

Fig. 63.

IMPORTS AND EXPORTS OF AIR

in millibars, thousandths of a ton per square foot, or tons per square dekametre

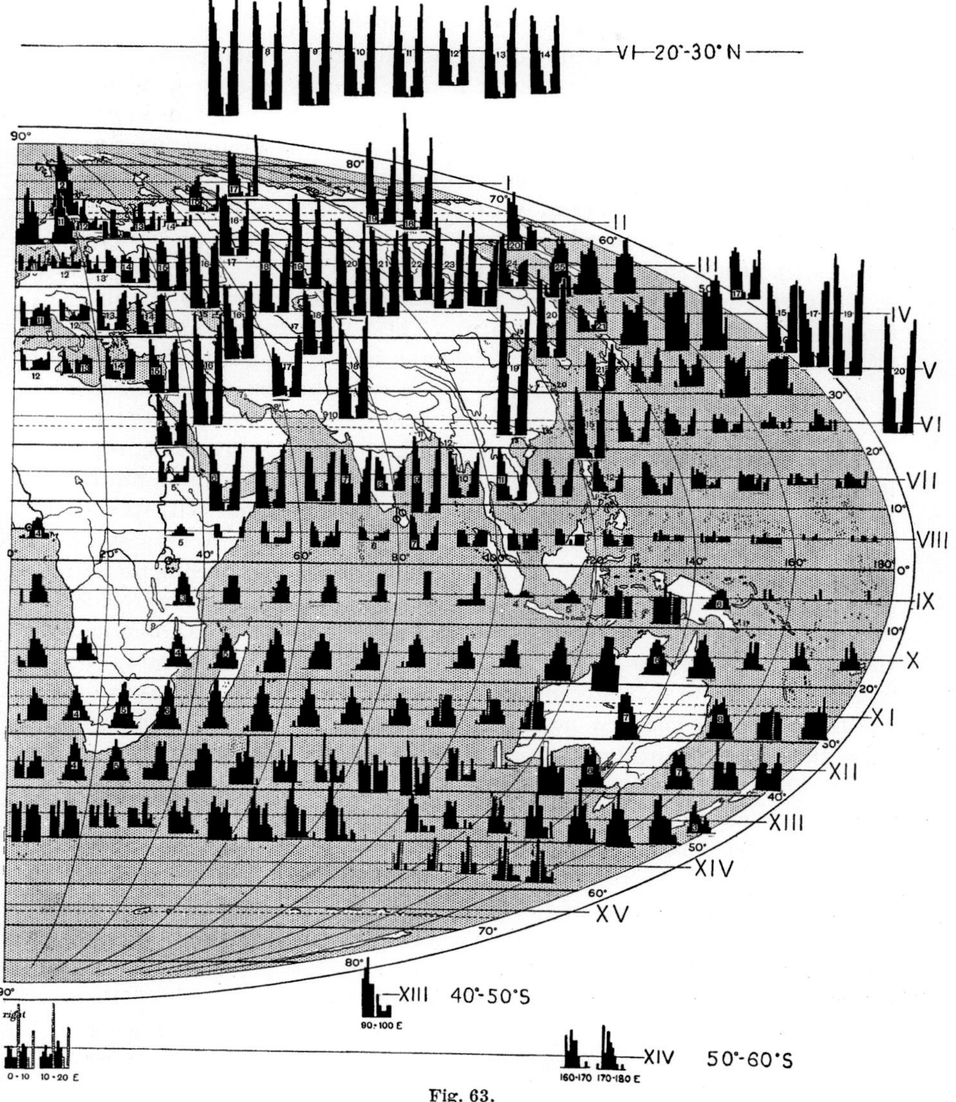

Fig. 63.

I2-2

Temperature over the globe

We continue our short story of the introduction to the drama of the world's weather as represented in the book of records by a series of step-diagrams (fig. 64) for some hundreds of places on land or sea on lines similar to those for rainfall and pressure. They show the normal maximum temperature of the day and the normal minimum for the day in each of the twelve months of the year. With the understanding that the maximum temperature is reached sometime during the 24 hours, generally in the day, and the minimum temperature likewise is reached, but generally in the night, we may regard the temperatures so marked as the day-temperature and the night-temperature. These two are indicated by the top and the bottom of a little black column for each month, and as the temperature gradually rises month by month the twelve little black columns for a station in the northern hemisphere make an arch; the external top of the arch is the normal maximum of the warmest summer month, and the lowest point of the abutment the normal minimum for the coldest winter month. The length of the black column tells its own little story about the range of the thermometer between night and day for each month. Long black columns like those for stations in Siberia and central Africa mean great difference between day and night; very short ones like those for the sea mean nearly uniform temperature throughout the 24 hours. A column higher than its neighbour on the left shows the rise of temperature with advancing summer, sometimes so rapid as to lift the minimum for the new month above the maximum of the old and leave the several monthly columns stranded as it were in the air. That is obviously the case over the sea where the range between day and night is never much more than a single degree and perhaps generally less though the seasonal variation is considerable.

The step-diagrams, diurnal and seasonal, of the change of temperature are set out in a rough map of the globe on what is called Mercator's projection, a method of mapping that is a favourite with seamen because it keeps the directions always "true". Up and down is real north and south, left and right real west and east. Meridians are all parallel and keep their distance apart across the map, north and south; and latitude lines are all west and east at right angles to the meridians.

Our map is not unlike the shadow which would be thrown upon the walls of a round tower by a lamp at the centre of a globe of the same diameter as the tower. One can understand that the shadow of the part near the equator would be a fair representation of the outlines on the globe; but as the part shadowed got more and more nearly over the shadowing lamp the step of ten degrees would throw a longer and longer shadow and no tower could be high enough to show the pole or the last few degrees of a meridian. That part of the globe is of no interest to commerce at present so that Mercator does not trouble about its not being represented on the map.

We have not hitherto used the popular Mercator, but now its expansion of the scale as one goes north or south from the equator is useful to us because the high latitudes are the parts of the earth where the seasonal temperature range is great; the enlargement of the scale accommodates our diagrams.

The whole chart is divided into zones of ten degrees of latitude by horizontal lines drawn across it. The scale of latitude is shown at the sides. The same scale of temperature is used for all the diagrams; in the original 1° C or 1 tt is represented by 1 mm and the reduction is to one quarter.

In order to enable us to compare the temperatures of the different places and for different months a common system is used for all the diagrams in a ten-degree zone. For latitudes between 40° N and 40° S the top bounding line of the zone is taken as 310 tt, 37° C, that is, in fact, within a fraction of a degree the normal temperature of the human body, below that is a cross-mark to indicate 300 tt, about 81° F, the temperature of a hot summer's day, and then 290 tt, 17° C, the restful temperature at which one can sit in a room without a fire. For places north of 40° N, or south of 40° S the top bounding line is 300 tt, 27° C, and lines are drawn in like manner for intervals of 10 tt below that temperature. A line is marked for 270 tt, three degrees below the freezing-point of water 273 tt, 0° C. In the most northerly zone of all the lowest line is for 220 tt, about – 60° F, over ninety degrees of frost.

Let us use the occasion to quote a number of temperatures that may be of interest.

TEMPERATURES OF THE DAY AND OF THE NIGHT

Pacific hemisphere.

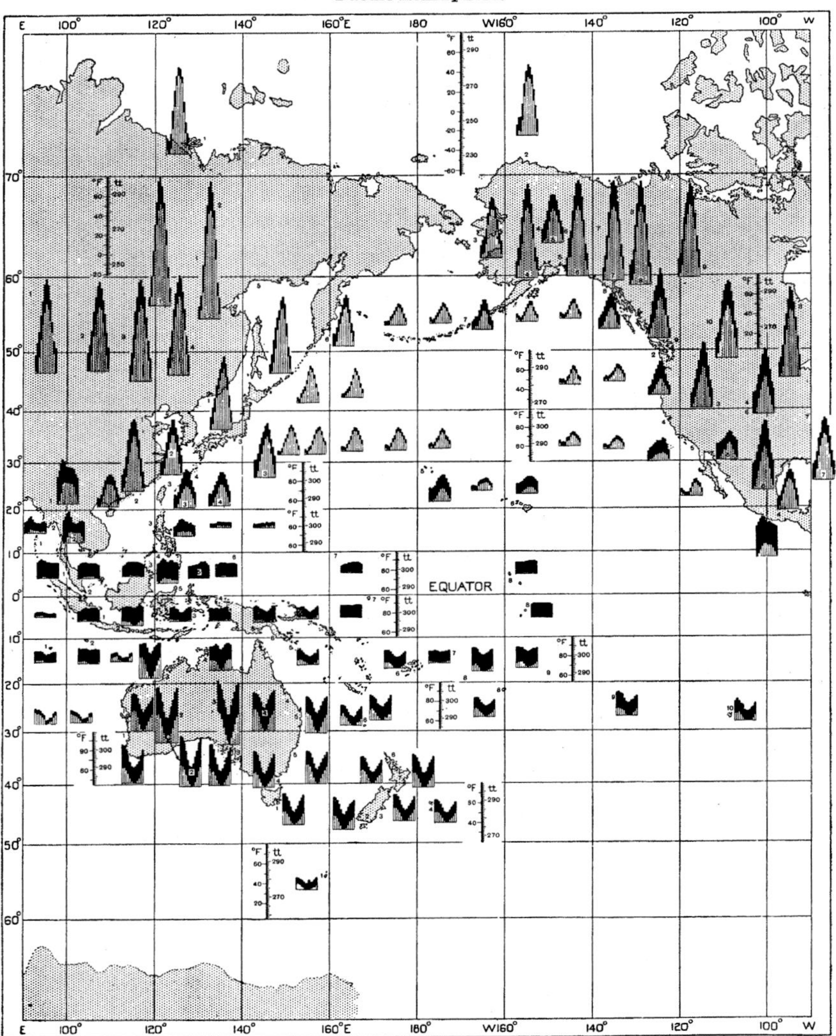

Fig. 64. Monthly normals of maximum and minimum temperatures at 198 stations on land and 40 five-degree squares at sea. Two specimens of the diagrams on a larger scale are given in fig. 61.

THE WEATHER'S GUIDE TO THE CHOICE OF A HOME

Atlantic hemisphere.

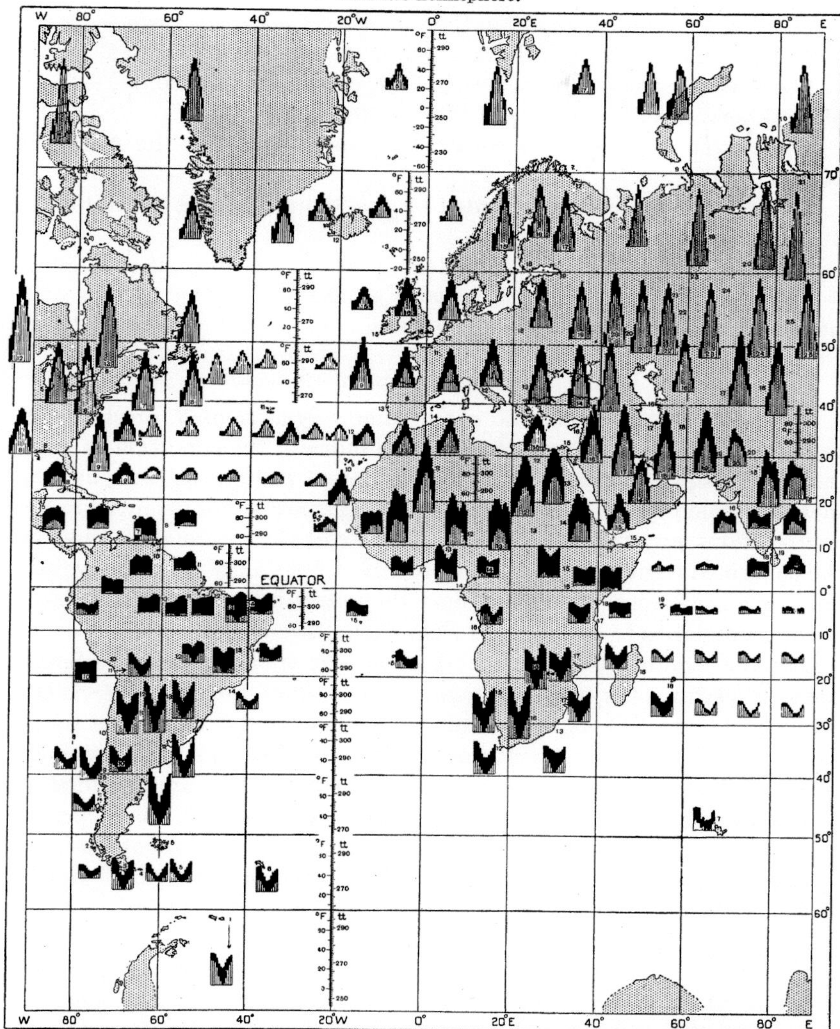

The normal daily maximum and minimum for each month at the several stations are indicated by the heights of the top and bottom of each black step of the diagrams for each station, as shown by the graduations of the temperature-scales marked within the ten-degree belt to which the station belongs.

The highest temperatures "in the shade" at the earth's surface, 329 tt, 133° F at Death Valley, California, 330·7 tt, 136° F in Tripoli.

The lowest temperature recorded at the earth's surface, 205 tt, – 90° F at Verkhoiansk in the middle of Siberia. (See p. 300.)

The lowest temperature recorded by Captain Scott in the Antarctic 239 tt, – 30° F.

The lowest temperature recorded by the Mount Everest expedition in 1924, 254 tt, – 2° F.

The lowest temperature recorded in the free atmosphere 180 tt, – 135° F over Java, near the equator in the Dutch East Indies at a height of 16 kilometres or about 10 miles.

Wind. The variableness of its habits

We should like to treat wind in the same way as rainfall, pressure and temperature, using the new wind-star to indicate the condition of wind in different squares of the ocean and at different stations on land; but the material which would be available for that purpose is not arranged in the compendious way that we suggest, and to edit the information to show the combined results would be the work of at least a life-time. So we can only give some specimens which include interesting contrasts for places in the same latitude and divided from each other only by sea.

Our first example tells something about the trade-wind of the South Atlantic ocean, everywhere recognised as one of the most persistent winds of the world. It carries air to the north-west along the south African coast and turns westward as it nears the equator passing St Helena in 16° S on its way. The roses which we present are intended to give some idea of what happens in the further stages of the perpetual circulation. Fig. 65 shows the normal winds for each month over a square in the sea in the neighbourhood of St Helena 15° to 20° S and 5° to 10° W. The directions from SE and ESE account for nearly all the winds without much variation from month to month. There are hardly any calms, the average is 1 per cent of the observations with 2 per cent at the northern midsummer. On the left is the corresponding diagram for a square 25° further west, not far from the coast of Brazil and there East shows to greatest advantage; but observations range round the compass from south through east to north-west, and the northerly components **ENE** and **NE** get their maximum in the southern summer while the southerly components wait their turn for the southern winter. If we pursued the matter further we might find that in the southern summer quite an appreciable quantity of trade-wind air prefers to turn southward along the Brazilian coast rather than continue to the equator and the West Indies.

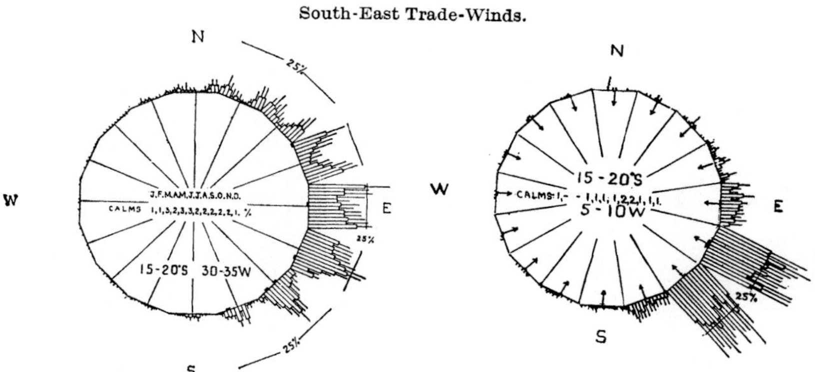

Fig. 65. Wind-roses from a zone of the SE trade-wind of the South Atlantic Ocean. On the right for the neighbourhood of St Helena, where arrows mark the June-July transition. On the left for the same zone east of the Brazilian coast, where lines to the centre mark the June-July transition.

The line of 25 % of the total number of observations *in each month* is indicated.

By way of contrast with the regularity of the trade-winds we may offer the comparison between winds over the sea to the west and the east of the English Channel (fig. 66); and here we must notice a difficulty that arises from the fact that the information comes to us distributed over sixteen points of the compass and we should prefer that it had been over eight. It is awkward to find a difficulty about knowing more than one wants, but it has to be faced. The rose on the left for the sea near the Scilly Isles reproduces the same information that we gave in fig. 55 but a kind friend has met the difficulty for us and somehow congested the sixteen points into eight. There is a great deal of information for those who seek. One of the features is the stressing of northerly and northeasterly winds in March and April, of easterly winds in the spring and autumn, north-westerly winds in September and south-westerly or westerly winds in January.

The companion picture for the winds of the southern North Sea reduces the distribution from sixteen to eight in a different way. It leaves the

Winds of the temperate zone.

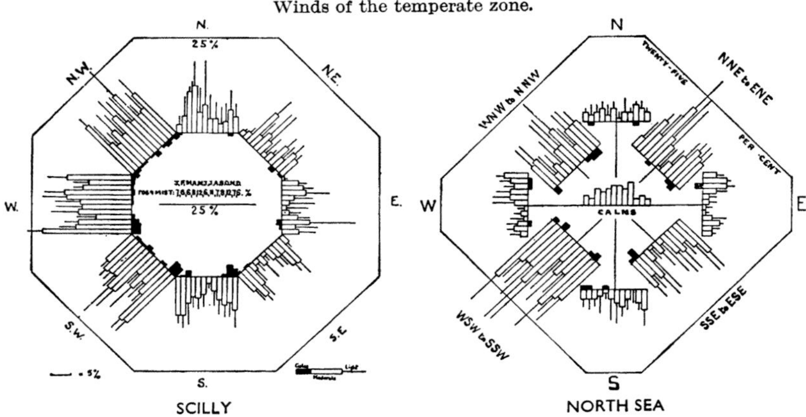

Fig. 66. Wind-roses for the western region of the English Channel and the extension of the Channel eastward into the North Sea.

The rose for Scilly on the left—the first of its kind—carries the explanation of the method. The information, originally for 16 parts, is reduced to 8 by division of the alternate parts between its neighbours.

The rose for the southern North Sea on the right has information for 16 parts reduced to 8 by grouping NNE, NE and ENE in a single wide angle, and so on. Lines towards the centre mark the June-July transition.

winds from the cardinal points in their original seclusion but groups together winds from NNE, NE and ENE in a single compartment as north-east, and so on for ESE, SE and SSE, and then of course the angle of the rose from which that compartment is furnished is three times that assigned for the cardinal points. There we notice NE winds in April, May and June, diminishing in July and August and getting up again in autumn, but frequent winds from SSW, SW and WSW all through the year. Those and the north-easters in April and May are the only winds that reach the 25 per cent line.

Another example (fig. 67) takes us back to the regular winds of the globe, in this case the monsoon of the Indian Ocean not far from the Persian Gulf. The winds are compacted into a rose of eight points by the plan of aggregating those for the three compartments between cardinal points, and the result shows the great prevalence between May and September of winds from the south-western points, and from the north-eastern points from October to March. The transition from one to

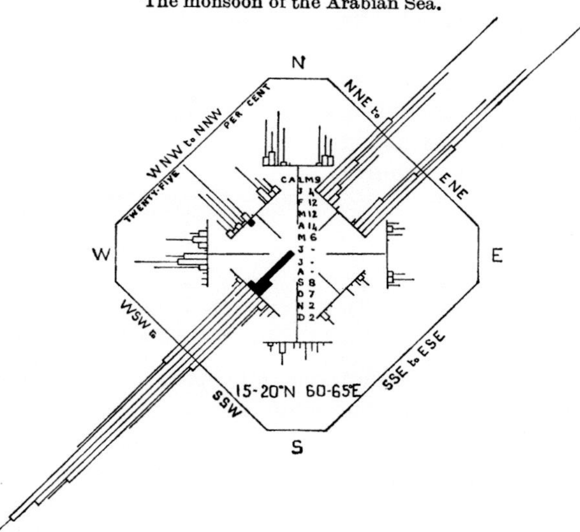

Fig. 67. Rose for the monsoon region of the Arabian Sea. Information for 16 parts reduced to 8 by grouping NNE, NE and ENE in one large angle, and similarly ESE, SE and SSE, and so on. Lines towards the centre mark the June-July transition.

the other is rapid and in that region there are gales in the south-western compartment.

A similar method of collecting the information about winds can be applied to the observations of winds in the upper air derived from pilot balloons[1].

[1] "A new sort of wind-rose." *Q. J. Roy. Meteor. Soc.* vol. LIX, 1933, p. 43.

In contrast with the behaviour of the winds of English waters, about 52° N (fig. 66), let us glance at a star for 55° to 60° S off the south of Cape Horn. Notice how the black ends indicate the relative importance of gales in that region, and the frequency and intensity of winds from the western side of the compass. It should be noticed that in this as in previous examples lines to the centre are given as an aid to differentiating the seasons in place of the indentation shown in fig. 55.

Rounding the Horn beyond the Roaring Forties.

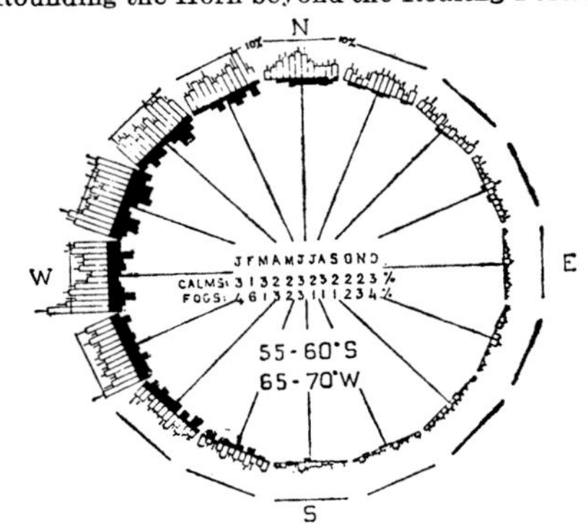

Fig. 68. Sixteen-part star. For each month's observations the 10 % line is marked.

APPLICATION OF THE SCORE TO THE FORMULATION OF THE SEQUENCE OF OUR WEATHER. WEATHER-FORECASTS

And now having set out the methods of keeping a watch on the pageant of weather and of recording the world's experience, we must go on to consider the use that may be made of the observations. We have a preliminary chapter with the title of The Chorus on the rhythmic aspects of our experience and their relation to days and seasons and other recognised periods, which may give us some idea of the sequence to be expected "as a rule" from the latent rhythms which are disclosed, before we venture upon the tremendous subject of the use of the weather-charts of to-day to foretell the experience of to-morrow.

In Britain the practice of forecasting from weather-maps was regularised in 1879 after a spontaneous effort on the part of Admiral FitzRoy at the Meteorological Office of the Board of Trade. It followed a desultory course before the war of 1914–18 but the development of aviation in the present century has resulted in enormous demands for information as to the weather of the present time and the prospects of future weather at any point of the various airways which at varying levels now lead all over the globe.

With this responsibility in view, in 1919 the Meteorological Office was attached to the Air Ministry.

The change in the official attitude towards weather-study is perhaps best expressed by saying that in 1905 the refusal of the Treasury to increase the grant for weather-study beyond the long-established figure of £15,300 led to the withdrawal of the Royal Society from responsibility for its management. In the expenditure on weather-study in the year 1936–7 under the management of the Air Ministry there is shown in the Report for that year a separate item of £16,583 in respect of Telegrams, Telephones, Subventions to reporting stations and miscellaneous charges, with a total net expenditure exceeding £208,000, and for the year 1937–8 the expenditure will be much larger as provision is made for the transmission of reports by teleprinter at any time of the day or night from a large number of stations so that flying may have the necessary security.

"From the meteorological radio-station of the Air Ministry at Borough Hill there is now a service of broadcast facts beginning at 7.15 a.m. and continuing hourly until 12.15 a.m. midnight on Mondays to Fridays, and until 6.15 p.m. on Saturdays and Sundays. The daily numbers of facts broadcast are respectively 8000 and 6400."

The modern investigation requires data from the upper air. A separate section of the Daily Weather Report is now devoted to the results, wherein special information from ships at sea is recorded.

As a matter of history we are retaining the division of our account of the methods of forecasting into two sections with the sub-titles of the Barometer Solo and the Norwegian Duet, and here we may note that the development of the second method, the method of fronts, has become acclimatised throughout the world and a special section of the Daily Weather Report is devoted to showing the fronts in the neighbourhood of the British Isles at 1h, 7h, 13h and 18h of Greenwich Mean Time.

There is another application of weather study and that is to trace the influence of weather upon growth. In that connexion daylight is one of the governing considerations and has to be regarded in association with the measurement of temperature, rainfall and sunshine. For that purpose the use of the week as a time-unit deserves further consideration as a development of the ideas of the Weekly Weather Report which, as set out in Chapter I, has lapsed to an annual issue. In this edition accordingly a supplement has been added with the heading "Chapter and Verse for Weather in relation to Agriculture", and appears as Chapter VII.

The supplement includes the information which was given at the end of Chapter III in previous issues under the heading "Aggregates for comparison with agriculture's natural integrals".

CHAPTER IV

THE CHORUS. RHYTHMIC ASPECTS OF THE RECORDS

Here will we sit and let the sounds of music
Creep in our ears.

LORENZO in *The Merchant of Venice*

THE UNIVERSALITY OF RHYTHM

Fig. 69. Waves and wavelets on the sea in a gale.

"A photograph taken when we were hove-to in a strong gale, force 9, in the bay of Biscay, December 21st, 1911, showing another ship hove-to riding on the crest of a wave between 30 and 40 feet high. The next following wave can be seen and also the very small waves made, as I suppose, also by the wind on the backs of the swift-moving waves; these were travelling at about 40 miles per hour. They were flatter than the waves of a moderate gale although higher.

"The scene was striking, but the most striking part was the movement, and part of that appearance was the difference between the swift rush of the great waves and the slow motion of the small ones. All that disappears in a stationary photograph and has to be explained if people are not to suppose that the tameness of such a picture is the fault of [the wave-maker or] the photographer."

DR VAUGHAN CORNISH.

In the maps displayed in figs. 62–64 the average of the measurements for each month derived from observations for a long series of years was represented by a little column and the twelve columns side by side showed what we may call the normal variation within the year at each station.

There is an obvious rhythm about the little diagrams; rainfall is highest in summer in the interior of the continents and in winter at the coast stations, the sequence from one month to another is regular and sometimes with little variation within the year; sometimes, as in the diagram for Cherrapunji, it shows so much variation that a special place has to be found for the diagram.

Rhythm represented by step-diagrams of quite similar character is shown in the new views of pressure and temperature, so that speaking of normals we may claim that the seasonal change in all the elements is obviously rhythmic. Judged by their performance on the average, the elements keep good time within the year. It is hardly necessary to say that one year is not exactly like another just as one bar of a movement is not exactly like the others; but in both the time-rhythm is kept. There is regularity as well as irregularity in the rhythm of the drama.

This idea of rhythmic change is to be found in all modern ideas of the universe and if we wished to point an analogy with the kind of regularity in irregularity which is so characteristic of the weather we should ask the reader to look at any picture of the sea: so in fig. 69 we have one to remind us of the familiar experience. There are ripples and disturbances but under it all the regularity of advancing waves.

Rhythm of temperature at close quarters

There is rhythmic change in the day's record of temperature and humidity and less prominently but still recognisable in the records of

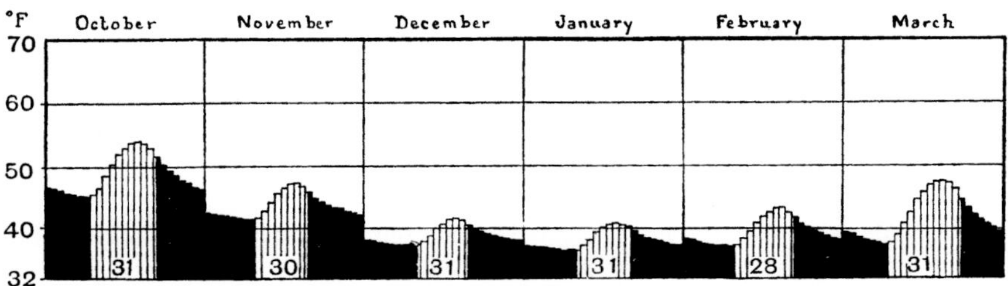

Fig. 70. The diurnal variation of temperature at Richmond (Kew Observatory) in each of the twelve months. The temperature is in degrees of the Fahrenheit scale; horizontal lines are drawn for each 10° above 40° F.

pressure at least as a half-day rhythm. To exhibit another feature of the analogy let us look at the diurnal variations of temperature at the Kew Observatory in the successive months based upon the observations of 25 years (fig. 70). In the first compartment under the heading October are 24 thin columns representing the normal temperature at each of the 24 hours in that month and so on, for November, December and then for January to September. As the cycle is complete in twelve months the beginning is at our discretion and the month of October is selected in order to exhibit the cycle of the seasons as a descent from autumn to winter, a recovery to spring continued to the summer maxima of July and August, and then the descent to autumn, a succession of waves combining to form a ground swell.

The columns belonging to the night hours are black, those for the day hours show white. To an observer at the left-hand end, watching the future in the shapes of the curves in front of him, the waving sea would seem a very natural analogy.

Our diagrams are of normals for the separate hours of the days within the month. They are, in fact, a different way of expressing the same figures that would be used in an isopleth diagram for Chap. III. If instead of the repetition of a mean October diagram thirty-one times (once for each of the thirty-one days of the month), we can picture to ourselves a sequence gradually changing with the passage of the days within the month, the illusion of the analogy with waves would be still more complete. The drop from the normal temperature of the last hour of October

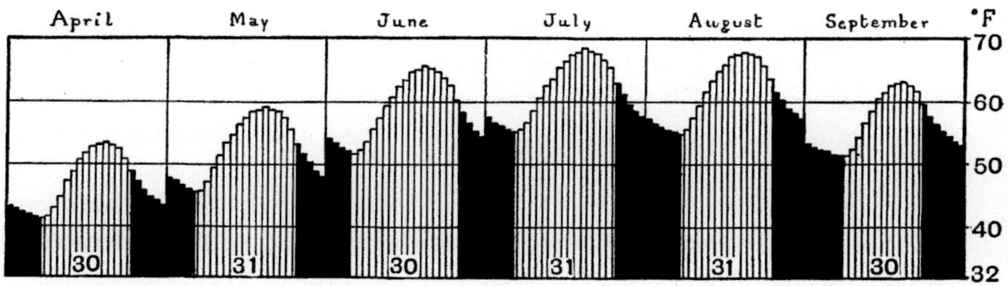

In each monthly compartment are 24 columns one for each hour of the day. The columns which are black are for the night hours, those which are white are for the daylight hours.
The lowest line represents 32° F the temperature of the freezing-point.

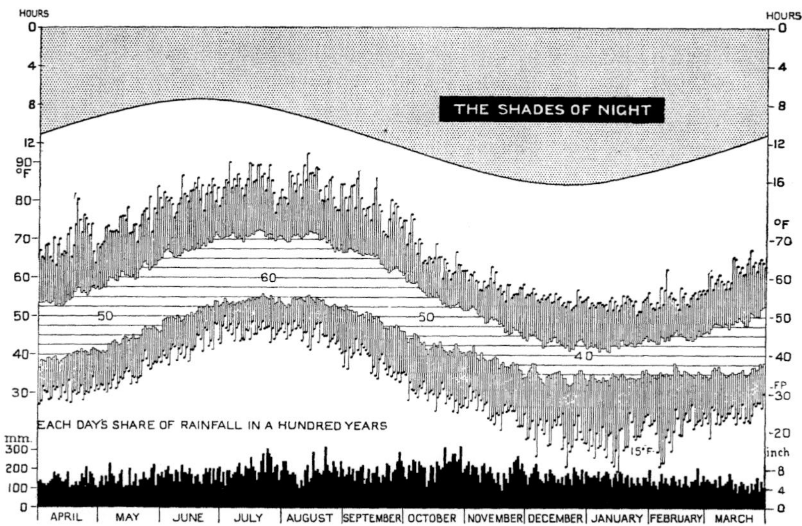

Fig. 71. In the middle the run of mean maximum and minimum temperatures at Richmond (Kew Observatory) with the extremes recorded for each day in the 30 years 1871–1900, and at the top a curve showing the length of the night. Between the lines of maximum and minimum, horizontal lines are ruled for steps of 2½ degrees of temperature, and the steps of 10° are marked.

At the bottom is the total rainfall for each day of the year in the London district in the century 1829 to 1928.

to the first hour of November is, in reality, a gradual change; but the range of the drop and the corresponding jump between April and May, speak for themselves as the indications which monthly means give of the gradual evolution of the daily range of temperature with the shortening or the lengthening of the hours of daylight.

Pressure and rainfall might be treated in a similar manner. For the average of a long series a rhythm would emerge; but irregularity would be more conspicuous than regularity in the daily curves.

Having reminded ourselves by illustration of the rhythmic character of change of temperature within the 24 hours, let us go on to represent a special example of the normal rhythm of the year at Kew Observatory (fig. 71). Here the day's record is compacted into the width of a line and the diagram shows the normal steps of extremes of temperature, maximum and minimum, day by day throughout the average year; and ordinarily, between those limits, the temperature at any moment may be expected to lie. The bravura of the thermometer in past experience is also shown by a highest note and a lowest note in the record for each day.

Here again, in the curves of mean maximum and mean minimum, we see a real rhythm for the year asserting itself, but rippled with slight variations like the waves of the sea. Along the rippled line we can obviously draw a smooth curve which will give us the underlying seasonal rhythm of the temperature for the observatory. But the range of the excursions of the thermometer above the normal maximum and below the normal minimum show that the mean is not by any means a full substitute for the facts.

Underneath the diagram of temperature is an inconspicuous representation of the fall of rain in the London district for each day of the year, compiled from the daily records of 100 years. This is a very real diagram because the rain-column magnified thirty times gives the actual depth of water accumulated on the date.

It gives us a diagram in which it is not by any means easy to see rhythm. There is a sort of suggestion of rhythm if one groups all the days of each month; October is a little wetter than April, and July has a little maximum of its own; but that will not entitle us to say that next October will be wetter than last April or the one next to come.

Rhythm and intrusion in the original record

Recognising, as we must, that the range of the bravura makes the normal curve of maximum and minimum, and still more clearly the curve of mean value, an inadequate expression of experience, we pursue the quest of curiosity a step further; the mean diagram for 30 years or 100 years is made out by combining the observations for each of the years, grouping them according to days and doing some kind of violence to the extra days of leap years. Let us look at the records for an individual year (fig. 72), not indeed one of those which have been included to give the normals; but there is no reason to think it belongs to any separate species or category.

Daily Weather at Richmond [Kew Observatory]

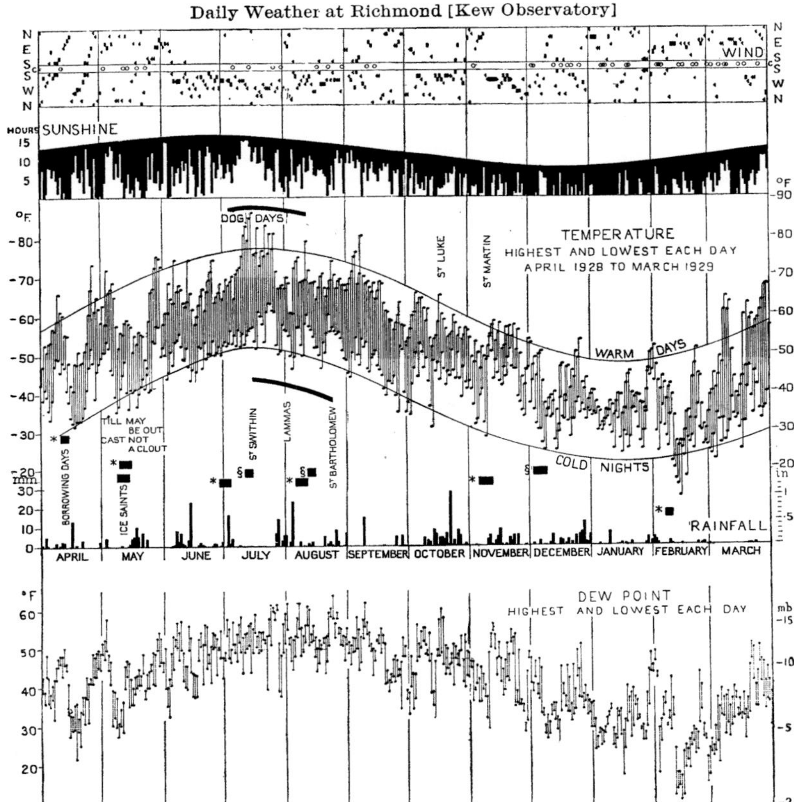

Fig. 72. Observations of wind-direction and force, sunshine, maximum and minimum temperature and rainfall at Kew Observatory for each day of the year April 1928 to March 1929, taken from the Observatories Year Book of the Meteorological Office, with the maximum and minimum of dew-point for each day taken from the observations at 1h, 7h, 13h and 18h reported in the Daily Weather Report, from results communicated by the Observatory for the purpose of this diagram.

Vertical lines are drawn on the diagram to show the separation of the months.

The wind-direction observed is indicated by the position of the symbol according to the orientation marks at the sides. The double line within which appears the symbol 0 (calm) is for winds less than 1 metre per second. Winds in the eastern quadrants are marked above the calm belt, in the western quadrants, beneath. The symbol ◀ means wind less than 5 metres per second, ■ between 5 and 10, ▶ greater than 10.

The black blocks marked * indicate the periods suggested by Dr Buchan as notably cold spells for Scotland and those marked §, on the other hand, the periods of his hot spells.

The year chosen is that which ran from 1 April 1928 to 31 March 1929, because it showed an unusually cold spell of weather in February. Here, of course, we have no curves of "mean values" between which the observations for any particular year may be expected to lie. We have only the actual observation of maximum and minimum for each day between which the readings of the thermometer have actually ranged; they give the extreme notes of the daily period, which in the figure are connected by a vertical line to show that they belong together.

Underneath again is the rainfall record showing the days on which rain fell and how much.

Here, again, we see an underlying rhythm in the curve for temperature though the individual notes show conspicuous irregularity; but there is no rhythm at all in the figures for rainfall. We could draw a curve for temperature that would make a sort of ideal of warm days or another that would make a sort of ideal of cold days, and similarly for warm nights and cold nights; and these ideal curves would obviously be subject to intrusion not so much by individual exceptional days as by groups of days, either exceptionally warm or exceptionally cold, that may be said to invade the curve and hold the control for a week, more or less, at a time.

So we must recognise that in thinking about our daily weather it is rather idle to have special regard to the rhythm of a general mean value. The maximum and minimum for any day are more pertinent than the mean; the mean curve for the year is only the paved pathway of approach to the realities of nature. There is a sort of change over, from a warm ideal to a cold one and back again to the original; and so on for ever. Meteorology cannot be content with the mean; it must seek for the ideals and the disturbances which invade them from the one side or the other. If consideration of space in the volume of the records demands a limitation of the information, the extremes for the day or the week are preferable to the mean. With the extremes a reader could not go far wrong in guessing the mean; but with a mean alone a guess at the extremes is not worth while.

Thus we can regard the sequence of temperature, which is perhaps the best index of changes in weather, as made up of a rhythmical sequence with interruptions due to some cause which we must seek in a future chapter and which, meanwhile, we will call intrusions.

The year 1928–9 was chosen for the diagram of fig. 72 because of the remarkable cold intrusion in February which is shown in the diagram to be an intrusion in duplicate, with a warmer spell between the two descents. A similar intrusion occurred in February 1932 accompanying a notable drought, and another in 1933 accompanied by snow and rain.

Some of the intrusions are so well known that their types have acquired special names with a reputation either for cold or warmth. We have marked on the diagram the borrowing days followed in 1928 by a fine specimen of the "blackthorn winter"—another occurred in the middle of April 1931; the Ice Saints SS. Mamertius, Pancras, Gervais, 11th, 12th, and 13th days of May; St Luke's summer a fine spell about the third week of October and St Martin's summer a similar return to the best of warm conditions about the second week of November. In order to complete the survey of weather conditions daily information about dew-point sunshine and wind have been added to the diagram.

Dr Alexander Buchan has assigned dates for cold spells and warm spells in the sequence of weather in Edinburgh—the dates of these also are indicated on the picture; but we can hardly expect London to imitate Edinburgh successfully.

We have chosen temperature as the element which will best express the analogy of the sea with the weather, and let the sounds of music creep in our ears, for the reason that its note is almost equally strong whether it is the day's record or the year's record that we wish to listen to; the combination of the two in fig. 70 is indeed a lesson in harmony. For the frontier provinces of middle latitudes where air coming from the frozen north or east is frequently in conflict with the wind which brings our temperate climate from the south and west, temperature may be the best index of the weather. We have included in our survey a specimen of the conflict in temperature with its varying fortunes (fig. 72), by taking also into account the intrusions upon the regularity of an ideal climate. If we were content with a more monotonous melody we could find in pressure a very efficient performer.

Example from the Indian Ocean

There is a little island in the Indian Ocean which has staged a remarkable example of rhythm, more monotonous than anything we can expect in middle latitudes, and on an occasion selected for our instruction an intrusion of such appalling violence that any country may regard itself as fortunate in not being able to record the like. Fig. 73 carries six days of the record of a barograph at Cocos Island in the Indian Ocean in November 1909 which shows a singularly regular half-day oscillation from the 23rd to 26th, and the intrusion of the depression of a hurricane about

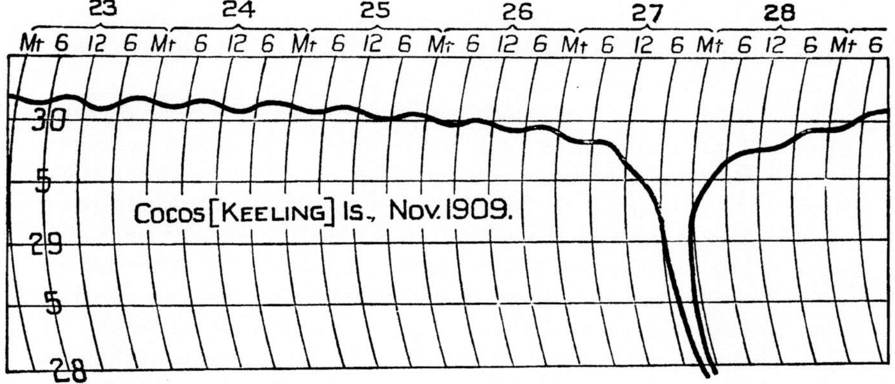

Fig. 73. The barograph record of a hurricane in the Indian Ocean in November 1909.

midnight on the 27th so deep that the paper could not register its extremity; and yet we follow the hum of the regular oscillation while the pen is plunging into the abyss, to be resumed "da capo" when the intrusive hurricane has moved elsewhere.

Some thought about this diagram may perhaps convey a useful suggestion about the absence of any rhythm in diagrams of daily rainfall. We may note that the range of the regular oscillation of pressure with its period of half a day is approximately a tenth of an inch on the barometer, and the intrusion which expresses itself otherwise as a hurricane causes a drop of more than 2 inches. Its occurrence is an isolated experience; if it were included in a series of averages its effect would be noticeable for many years. Pressure on twenty successive 23rds of November might be averaged and still be affected to the extent of a tenth of an inch. And there is no chance of obliterating its effect by a similar dislocation in the opposite direction—simply such behaviour does not happen. If we ask why the depression is natural but a corresponding elevation quite out of the course of nature we appeal to rainfall for an explanation. Very heavy rainfall is an accompaniment of the development of a hurricane; and indeed it is fair to say that when the elements for making a hurricane are within range it is the rainfall that takes control of the situation and accomplishes the task. There is a certain instability which attends the formation of rain and accounts for its revolutionary activity. It plays the big side drum in the weather's orchestra.

Thus rainfall seems to be associated with intrusions upon the rhythm; but not all intrusions are so conspicuous as the Cocos Island hurricane. We shall cite quite a number of different kinds which disturb what would otherwise be a tranquil rhythm; but if we are disposed to complain of that interference as one of the unkindnesses of nature we must remember that it is just the intrusion of consonants that makes the difference between the miracle of human speech and the plain song of the whistling wind. However vexatious they may be for the meteorologist seeking an easy explanation, meteorological intrusions are the language of the play of wind and weather. "Deep calling unto deep at the noise of their waterspouts": mountains of the far north and of the far south talking to the continents across the oceans.

RHYTHM AND REALITY

Attracted by the rhythmic appearance of the groups of means for each of the twelve months of the year for rainfall, pressure and temperature in our discussion of the score and the records, we have found some ground for regarding the sequence of maximum and minimum temperatures at Kew Observatory as a series of intrusions from the one side or the other while, nevertheless, the sequence exhibits an easily recognised rhythm for the day and the year.

Philosophers may argue as to whether the rhythms at which we have arrived are real or unreal. We would prefer not to have to argue about the twelve-hour rhythm of pressure in Cocos Island which has a peculiar claim for reality. Let us allow that no single day or year ever exhibited in the past the rhythm which we have drawn from normal values nor ever will in the future. And yet the idea of rhythm gives valuable information for those who wish to keep the way of the world in mind. That view receives strong support from the fact that very similar rhythms are arrived at for neighbouring places and for successive periods of time. There is perhaps a tendency to think that the longer the period the more real the rhythm, that a mean for 35 years is better than a mean for 10 and not so good as a mean for 100; but the contention will hardly bear examination by a meticulous philosopher. Not one of them can be called absolutely real and any one of them is sufficiently nearly real to serve as a guide to our experience.

So we propose to pursue the subject of the seasonal rhythms still further, to group together those which attain their maxima together and may be said to be keeping time, and to put into other classes those that attain a maximum at equinoxes instead of solstices or at some other phase of the earth's march round the sun, and so get some idea of other ways of the world than our own.

WEATHER'S SYMPHONY OF RUGGED RHYTHMS. SYNCHRONISM, SYNCOPATION AND INTRUSION

For our immediate purpose it is sufficient that when we gather together into separate compartments the records of the same month in successive years, the result generally shows a rhythmic sequence not unlike the sequence that would express the rise and fall of a boat or a buoy as the waves pass, each wave representing a year. A similar statement would generally be true if we grouped the observations according to the hours of the day, instead of the months of the year, in which case the passage of a wave would correspond with a day instead of a year. So a library of the world's weather provides us with the means of expressing the result of a succession of weather changes as a rhythmical sequence for what we call the average year or the average day, though the average year or average day may be one that never was exactly on land or sea.

The year and the day have been chosen because the revolution of the earth round the sun in its orbit and the rotation of the earth on its axis expose the atmosphere to rhythmical changes in the number of units of energy that comes from the sun as radiation; and it would be something in the nature of a miracle if so powerful an agency left no trace of the sequence of its changes upon the condition of the atmosphere. Anything on the earth which proved to be independent of day and night and of summer and winter would certainly deserve to be called supernatural.

The process of grouping and adding by which we have ascertained the diurnal or seasonal rhythm is effective for the purpose because it dilutes the influence of the intrusions that spoil the perfection of the rhythm. The figures which mark the extremes of each day in the diagram of the variation of temperature at Richmond have all been accounted for in the mean to which they are attached; but the excesses which they record have been combined with all the other values for the same day of the year and in a curve for 30 years the influence of each divergence from the mean value is reduced to one-thirtieth of its own bravura, and an occasional special effort may be compensated by failure to reach the normal on another occasion. We may, therefore, detect a rhythmical variation, which we will call real, in circumstances which for the individual records would seem very dubious.

The rhythm is expressed in our step-diagrams by successive steps leading up to a maximum, in one of the months or one of the hours, and then stepping down again to a minimum to come back to the original value. The order of the steps by which the sequence is disclosed depends upon the position of the first step which may be up or down. If it is up, the ascent will go up to the maximum, if down it will continue to the minimum. The simplest form of perfect rhythm is one in which the downward steps from the maximum to the minimum are just the reverse of the upward steps from the minimum to the maximum, and one of the essential features of the rhythm is the time of achievement of the maximum and the minimum. For temperature the maximum of the hours is generally at 2 or 3 o'clock in the afternoon, or in July or August among our months.

For that element the minimum does not generally show perfect symmetry; it occurs just before sunrise in the hours and in the months it hovers about January, February and March.

Now there are quite a number of things about the weather which in the monthly means of a number of years show an annual sequence with seasonal rhythm. They are certainly all associated in some way with the seasonal change in the solar radiation; but they make their own choice of the time at which they will show their greatest and least achievement.

We want some suitable word to express the behaviour of the different rhythms in this respect. Agreement as regards the month of maximum we may perhaps describe as synchronism, which means keeping time. For the delay or lag in the arrival at the maximum within the period or bar, we can borrow from music the use of the word syncopation, and we still retain the word intrusion for something that changes the key more or less abruptly without any permission from the observer, granted or asked. Syncopation represented by the lag of a month in temperature in relation to the longest day is clearly shown by the comparison of the daylight curve and the temperature curve in fig. 71.

An assortment of rhythms

The things about the weather that can be associated in this classification under the idea of rhythm are very numerous. Here is a whole series of them introduced by their diagrams. We begin with the total amount of energy received by the northern hemisphere from the sun and incorporate with that other rhythms which have their maxima and minima at the solstices, 21 June and 21 December, midsummer and mid-winter. Those of the next group have their maxima at the equinoxes 21 March and 24 September, spring and autumn, and a third group shows progressive syncopation in steps of two months in seasonal rainfall; the particulars of the representation are given by the side of the diagrams.

The year's diagram has been repeated in each case to suggest that the sequence is rhythmic and not simply a heap without relation to before or after. The reader will agree that the perfect repetition there indicated is just what will never happen in real life and yet it gives an ideal with which we can associate the intrusions which are characteristic of life.

I. Solstitial rhythm. Maxima in midsummer of the northern hemisphere

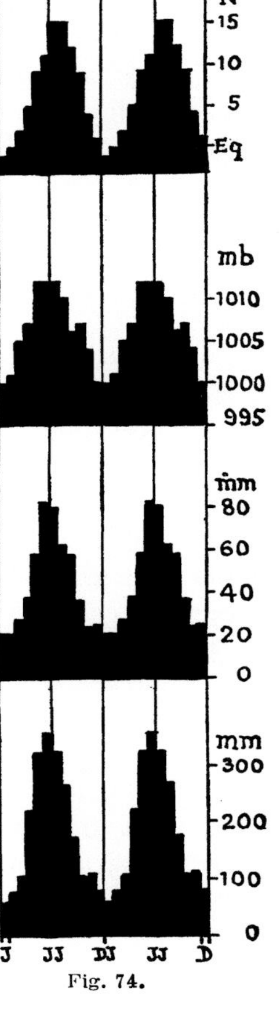

1. Solar radiation in the northern hemisphere. The numbers 1, 2, 3 indicate the scale of 1,000,000,000,000,000 kilowatt-hours per day. A kilowatt-hour is the unit for sale of electrical energy, costing generally something between ½*d*. and 8*d*.

2. The latitude of the belt of rain across equatorial Africa which moves up and down across the equator. The scale at the side shows the number of degrees north of the equator that the margin of the rain-belt has reached. See fig. 51, p. 156.

3. The pressure in the central portion of the area of low pressure which in monthly maps is shown over the Atlantic, south of Iceland and is bounded by the American and European coasts. See fig. 53 *a*, p. 160.

The scale at the side shows the pressure in millibars.

4. The incidence of rainfall in Winnipeg in the middle region of the Canadian prairie.

The height of the column shows the monthly rainfall in millimetres; 40 mm and 80 mm are marked.

5. The incidence of rainfall at Punta Galera, off the southern point of South America on the eastern side. The rainfall is very large; the scale in millimetres is marked. For three months of the year the fall exceeds 300 mm (12 inches) per month. Mean British rainfall is about 1000 mm per annum.

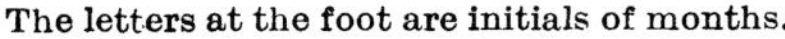
The letters at the foot are initials of months.

Fig. 74.

II. Solstitial rhythm. Maxima in midwinter of the northern hemisphere

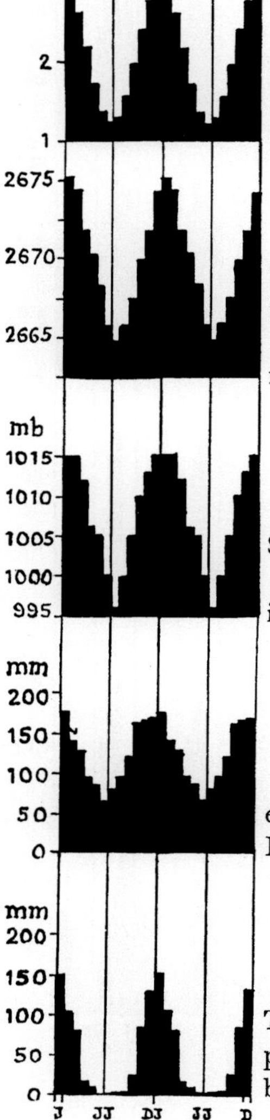

6. Solar radiation in the southern hemisphere. A little larger maximum than in the northern because the earth is nearer the sun in northern winter. Scale again 1,000,000,000,000,000 kilowatt-hours.

7. Total weight of air on the northern hemisphere. The scale shows billions of tons; 2670 billion tons is the normal load on the hemisphere.

The maximum in January indicates a syncopation in the rhythm of something like a month.

8. Pressure within the monsoon area of India. See figs. 53*a* and *b*, pp. 160, 161.

The figures against the columns show the pressure in millibars.

9. The rainfall at Thorshavn in the north-eastern part of the Atlantic between Shetland and Iceland. The figures indicate the scale in millimetres.

10. The rainfall at Bulawayo in South Africa. The figures show the scale in millimetres; there is practically no rain in winter, June to September, but the summer fall is strikingly rhythmic.

Fig. 75. The letters at the foot are initials of months.

III. Maxima in spring

Each year's diagram is duplicated.

11. The ebb and flow of air from the northern hemisphere across the equator. This time the unit is only a hundred thousand million tons; it requires ten of them to make a billion. The marking of the scale measuring from 26 as the datum-mark shows an ebb of about 26 units for March–Feb. and June–May, and return flow of like amount for Oct.–Sept. and Dec.–Nov.

12. The flow of air in the north-east trade, obtained from observations at sea.

The mean value is assessed at 4·9 metres per second, and the range from 1·1 above the mean to 1·6 below.

13. The number of tornadoes in Alabama in 132 years. A very peculiar kind of rhythm for a very peculiar phenomenon.

14. Rainfall at Meshed on the border of Persia. Spring is a very unusual time for rainfall maxima but the maximum of the Asiatic continent beginning with Meshed in March marches northward to take up a July position farther north.

15. Difference of sea- and air-temperature in the Atlantic 60 to 65° W, 35 to 40° N. The diagrams show the amount by which the temperature of the sea exceeds that of the overlying air. Another very unusual time for maximum associated with remarkable vagaries of February weather over the main steamer route across the ocean.

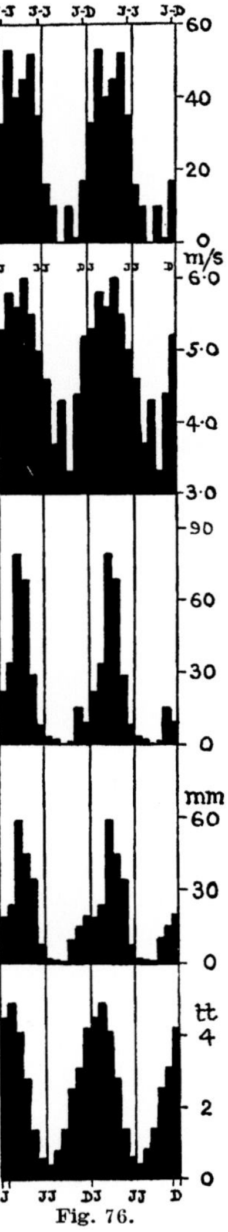

The letters at the top are initials to show monthly differences.

Fig. 76.

IV. Maxima in autumn

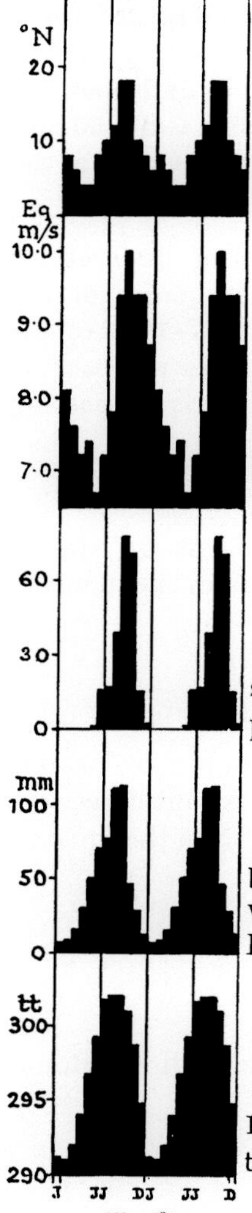

Fig. 77.

Each year's diagram is duplicated.

16. The northern limit of the region of doldrums of the Atlantic where the airs of the separate circulations of the two hemispheres are in juxtaposition. The scale is degrees north of the equator.

17. The strength of the south-east trade to be contrasted with that of the north-east on the opposite page. Not quite, however, on all fours as the figures are from an anemometer on St Helena, 2000 ft above sea-level.

18. West Indian storms in 37 years, remarkably similar to the curve for tornadoes in Alabama with the period of maximum changed by six months.

19. Rainfall at Vladivostok showing the end of the progress of the month of maximum across Asia, which begins at Meshed in March or at Robat in February.

20. Earth-temperature at Helwan at a depth of 1·15 metres (mean of $2\frac{1}{2}$ years) to be contrasted with the surface-temperature of the Atlantic opposite.

The letters at the foot are initials of months.

V. Progressive syncopation in seasonal rainfall

These diagrams have 2 mm to a month, but are not repeated for a second year. The initials of the months are at the foot.

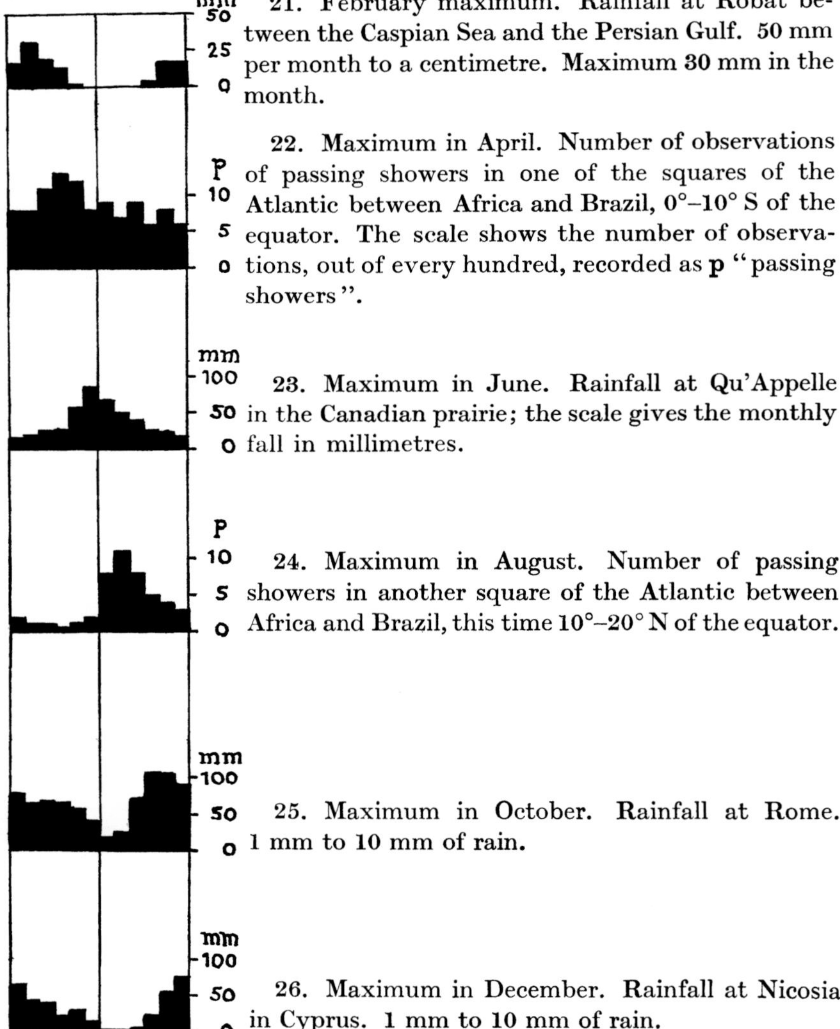

Fig. 78.

21. February maximum. Rainfall at Robat between the Caspian Sea and the Persian Gulf. 50 mm per month to a centimetre. Maximum 30 mm in the month.

22. Maximum in April. Number of observations of passing showers in one of the squares of the Atlantic between Africa and Brazil, 0°–10° S of the equator. The scale shows the number of observations, out of every hundred, recorded as **p** "passing showers".

23. Maximum in June. Rainfall at Qu'Appelle in the Canadian prairie; the scale gives the monthly fall in millimetres.

24. Maximum in August. Number of passing showers in another square of the Atlantic between Africa and Brazil, this time 10°–20° N of the equator.

25. Maximum in October. Rainfall at Rome. 1 mm to 10 mm of rain.

26. Maximum in December. Rainfall at Nicosia in Cyprus. 1 mm to 10 mm of rain.

Thus we are able to display twenty-six examples of rhythm apart from the regular series of rhythmical changes in temperature that are directly controlled by the sun's influence.

We can hardly fail in the long view to visualise an underlying rhythm in the movements of the atmosphere which produces the weather of the world; and yet experience warns us that days or weeks or perhaps even months may happen when the key of the rhythm is lost or unrecognisable.

Sometimes we can recognise definite intrusions and sometimes even venture to suggest a cause. Among notable intrusions within living memory are the wet summer of 1879, the eruption of Krakatoa in 1883, the cold spring of 1895, the very mild and stormy winter of 1898–9, the rainy season of 1903 attributed to the eruptions of La Souffrière and Mont Pelée, the warmth of 1911, the interruption of sunlight in 1912 by the eruption of Katmai, the drought of 1921, the cold February of 1929 and of the same month in 1932.

These are local memories; every locality has a store which is not shared with all the world.

DAYS, WEEKS, MONTHS AND YEARS

Perhaps this is the place to say a word about the grouping of observations according to periods. We have used already the hour, the day, the week, the month and the year.

The day and the year are natural periods and to each there corresponds a rhythm which can be elicited by combining the hourly observations of a great number of days, or the daily observations of a great number of years.

The week and the month are different. The month does, in fact, approximate to the period of the moon's changes from new moon to new moon; but that is not an exact number of days nor a simple fraction (an aliquot part, as we used to call it) of a year. It is about 29½ days and our twelve unequal months represent the effort of Julius Caesar and the substitution of his great nephew Augustus for a solution of the impossible task of dividing the number of days of the year into twelve equal parts.

To measure time something between the day and the year is so desirable as to be recognised as necessary. In the days when modern meteorology was beginning, the month of the Gregorian calendar was adopted as an intermediate period and at the time the revolving moon was regarded by many as controlling the weather; but in these days we are conscious that, in spite of the name, months do not correspond with moons and they are not of equal length. So in reality the calendar can be used to mark time but not to measure time. The period which finds a place in the social order is the week, and a question arises whether meteorology should adapt itself to the social order and use the week.

There is no obvious weekly rhythm or monthly rhythm of weather. And in fact weeks and months are only used now as arbitrary divisions of the year. Thus, the numerous step-diagrams of various features of the weather in this and the preceding chapter are not improved by the inequality of the lengths of the months, quite the contrary. The reasons for their continued use are first that they have been in use for very many years, and secondly, that twelve of them put together make up a year, whether it be one of 365 days or 366, and twelve is a convenient number for tables. If we use weeks of seven days we have fifty-two of them and one day or two days over in every year. For ordinary application of meteorology, the week is much more suitable than the month as a time unit. It gives a better shot at the duration of a spell of weather than a month.

The use of what we have called a daylight cycle in our illustrations of agricultural meteorology included in Chapter VII may suggest the method of using the week to the best advantage. Meanwhile it is not fair to say, without careful inquiry, that the day and year are the only rhythmic periods.

WEATHER PREDICTION BY LATENT RHYTHM

Of all the relationships of weather nothing is more apparent than the diurnal rhythm and the seasonal rhythm of temperature which, moreover, can be directly traced to their original source, namely the varying power of solar radiation in consequence in the one case of the rotation of the earth on its axis and in the other case of the orbit or path of the earth round the sun with, in each case, the inclination of the earth's axis to the plane of the orbit at $23\frac{1}{2}°$.

The normal rhythm is exhibited in the curve of normal diurnal variation of temperature in fig. 70 and in the diagrams of monthly variation in temperature for the air of various parts of the sea and land in fig. 64. There are places near the equator where the diurnal rhythm is the only survivor and on the other hand over the sea the diurnal rhythm has very little amplitude—the seasonal rhythm is the survivor.

These are the most conspicuous rhythms; but persevering persons are able to detect other rhythms in the sequence of values extending over a number of years. The method of procedure is by what is called harmonic analysis based on a theorem developed by the mathematician Fourier that any sequence of events can be represented by a series of rhythms beginning with one that has the full period of the sequence and combining with it one of half that period, a third and a fourth of it and so on, until adequate representation is achieved. For practical purposes only the rhythm of full period is used, combined with those of one-half, one-third, and one-quarter; but sometimes the fifth and the sixth are of importance.

The statement that any sequence whatever may be analysed into component rhythms by the process may seem to imply that a real rhythm may be found in every sequence; but analysis evidently loses its practical application unless the rhythm repeats itself in successive periods, as we may understand from the case of the seasonal rhythm which may show some slight variation in its details from year to year but will certainly assert itself in the mean values for the period of 35 years which we have called normal. So, if we wish to examine a rhythm for some other period than a year, for example, eleven years, another favourite, we may set out the values for each successive group of eleven years and taking the average of the sequence in the successive groups discover whether the mean values show something which is comparable or in any way like the normal seasonal rhythm.

In this manner with suitable modifications and developments various rhythms have been claimed as established. The eleven years, the average period of development and disappearance of spots on the sun's face, finds an echo in many earthly values, specially the level of water in Lake Victoria under the equator (fig. 79); but that was originally suggested

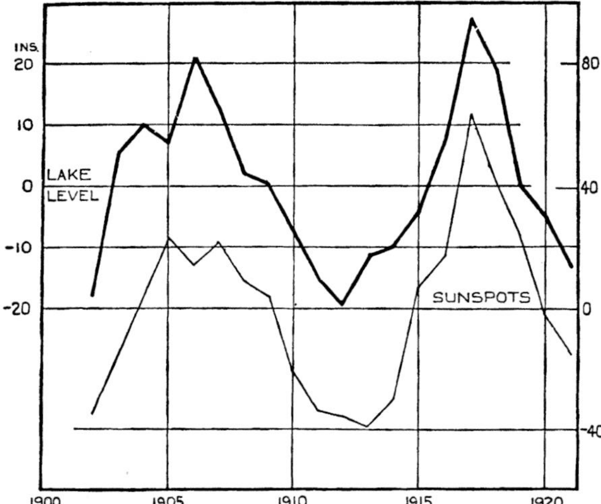

Fig. 79. The variations in the mean number of sunspots and the levels of Victoria Nyanza year by year from 1902 to 1921.

by the casual comparison of the actual graphs of the two varying elements.

Another attempt to use apparent correlation, based upon the similarity of curves of mean values (fig. 80) for the months, of the wind at St Helena and the rainfall in southern England ended in disaster. When public attention was called to the similarity a popular journal was convinced that strong wind at St Helena would mean rainy weather in London, and obtained information as to the winds direct from the island; but the result was very disappointing. The mean values in each may have a similar rhythm but the intrusions are different and we may almost say with confidence that it is the intrusions that count in rainfall.

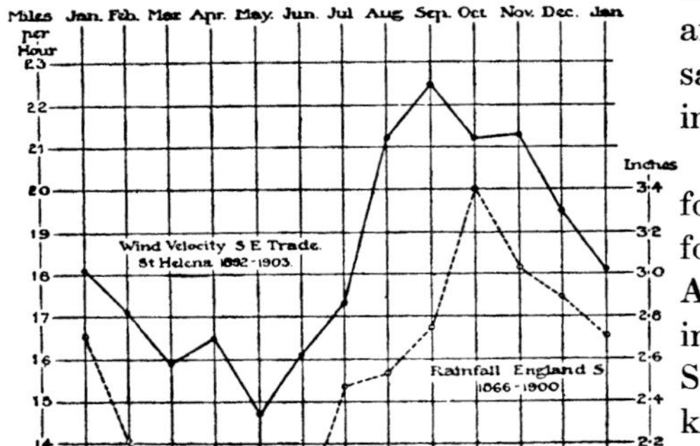

Fig. 80. Average seasonal variation of the velocity of the south-east trade-wind and of the rainfall in England S.

There is a three-year rhythm for rainfall at Oxford and also for pressure at Port Darwin in Australia, and a five-year rhythm in winds from cold quarters at Southport, and perhaps best known of all a 35-year rhythm, in weather, which according to Francis Bacon was already a subject of inquiry in the Low countries at the beginning of the seventeenth century. It was investigated in great detail for rivers and lakes, harvests and vintages all over Europe by a Swiss professor, E. Brückner.

In the same way hundreds of possible rhythms have been suggested as forming part of the great chorus of weather observations, and not infrequently one sees predictions based upon these rhythms quoted in newspapers as portending an imminent cold or wet, fine or dry season.

Here is an example as set out in a recent Manual of Meteorology.

"The lunar-solar cycle of 744 years has been invoked by the Abbé Gabriel. It combines 9202 synodic revolutions, 9946 tropical, 9986 draconitic, 9862 anomalistic, 40 revolutions of the ascending node of the lunar orbit and 67 periods of sunspots. It has harmonics of 372 years, 186 years. The last was relied upon for a prediction, made in the summer of 1925, of a cold winter to follow. The prediction was fulfilled in England by the occurrence of exceptionally cold weather in November, December and January. It must, however, be remarked that February, which is accounted as a winter month, brought the highest recorded temperature of that month for 154 years, and a spell of warm weather compared with which the first half of the month of May was winterly"!

Still more recently the Abbé Moreau has made public announcements of the weather of seasons to come.

Some day we may trace the causes of this recurrence of seasons assumed as a basis of long-period forecasting; but let us be warned by the experience with the diurnal and seasonal variations which any one may test for himself at any time. The rhythm shows without any doubt that for Kew the hour from 1.30 to 2.30 is the hottest hour of the day, and July is the hottest month in the year, January the coldest; but if on that basis I should predict that to-morrow 2 o'clock would be warmer than 4 o'clock I might be wrong; and equally so if I predicted that next July will be warmer than next September. I can rely upon it if it is an average of years that I use for checking my prediction but for a single year the prediction may be a harassing failure.

And so with the other rhythms, or periodicities as they are called. They may fairly be regarded as latent in the actual course of events extending over a hundred years; but for a single occurrence, this year or next year, they may be disappointing. So it is with the 3 years' rainfall at Oxford, the 5 years' cold winds at Southport, and the 35 years of the Brückner cycle; they are good enough to guide one's general examination of the sequence of events but not sufficiently dominant to tell us what will happen next winter.

Syncopation and Correlation

There is another method of representing a series of observations that enables one to form a general idea of the sequence of events. We have seen that in the seasonal sequence there may be a relation between two events, the one in time with the other or following it after a recognisable period.

Statisticians have developed a method of expressing these relations numerically by what is called correlation. Successive measures of two quantities, rainfall, temperature or pressure, may be arranged as deviations or differences of each from its own mean or normal value. The corresponding deviations from the mean are systematically compared by a numerical operation which gives what is called a correlation coefficient, or the fraction of the deviation of one of the quantities which is controlled by the deviation of the other. The coefficient 1 means that the one deviation is actually and rigorously proportional to the other, a coefficient 0 means that there is no relation at all between the two—they are varying quite independently; and a coefficient -1 means that the deviations are rigorously proportional but opposite. Coefficients $+1$ and -1 are very rare, practically unattainable while observations are liable to error. Generally the coefficients obtained are decimal fractions; a coefficient above ·6 has a good deal of meaning, below ·4 it may be quite delusive. All the figures are arrived at by drawing inferences from a long series of corresponding values and if we suppose a coefficient as large as ·8 or even ·9 it means that strict proportionality is broken on occasions by the disobedience of one or the other. So again while we may infer from a high coefficient that there is some relation between the two varying quantities of which in time we may find the physical reason, it would not be wise to publish an unguarded prophecy about the value of one of the variables for a particular year.

So what it comes to is this; we may use the predictions by harmonic analysis or correlation for suggesting to wise persons certain possibilities of relationship for closer investigation; but it would be quite unwise to publish the prediction in the press without so many *caveats* that the man-in-the-street would be disposed to write it all down as mere words and hedging.

So long as we are not fully cognisant of the action of the drama, or, as Clerk Maxwell expressed it, of the real "go" of things, such predictions are like the ancient Greek oracles or the prophecies of the Sibyls. It takes a wiser man to appreciate them at their real value than it does to make them.

The situation is by no means uncommon. We are accustomed to say that history repeats itself; and, on some occasions, the remark is sufficiently near the truth to be worth making. On the other hand it has also been said, and with as much truth, that repeating itself is precisely what history never does.

It is in fact precision, in history or in the study of weather, that makes the difference between the truth and untruth of the aphorism. We may agree, with the philosophers, that the same causes always produce the same effects; and, so long as we are dealing with the general aspects of the questions at issue, that similar causes produce similar effects. But for the student of weather as for the student of history the differences in cause or effect which are marked by precision are always important and sometimes more important than the apparent similarities.

It may, accordingly, be best to regard our rhythms and our co-efficients as poetic illustrations of the meaning of our facts and not as substitutes for them.

THE PERSONS OF THE DRAMA ARRANGED FOR THE BAROMETER-SOLO

Isobar—a line of equal pressure drawn on a sea-level map of the area under consideration.

Depression—an area within which the barometric pressure reduced to sea-level is less than that of its immediate environment, indicated on one or more of the isobars of the map by a local bending or bulging of the lines towards the region of higher pressure, and on the record of a barograph by a dip in the curve generally recovering in whole or in part within a day.

Cyclone—a depression of sufficient importance to be marked on the map by a series of closed isobars, generally drawn as oval curves. The centre of the smallest of the ovals is the centre of the cyclone. The winds circulate round the centre counter-clockwise in the northern hemisphere; clockwise in the southern.

Secondary depression—a depression marked by distortion of the isobars and the deflexion of the winds within the region of the isobars of a cyclone. The distortion may be a simple bend or a local centre may be indicated by a closed isobar.

Front of a cyclone—the region of the map identified by the extension of the area of the closed isobars in the direction in which the cyclone is travelling.

Rear of a cyclone—the last part of the area included within closed isobars behind the centre.

Trough—the line drawn across the cyclone on the map through the centre at right angles to the direction in which the cyclone is moving. The part of the line which is drawn to the south (south-west or south-east) of the centre is generally associated with sudden changes of pressure, wind, temperature and humidity but the part of the line from the centre to the north side is not so marked.

V-shaped depression—the trough-line is sometimes associated with a distortion of the isobars into V-shaped lines with the points of the V's marking the trough-line.

Discontinuity—changes in pressure, wind, temperature and humidity on the trough-line or elsewhere so abrupt as to be classed as discontinuous.

Anticyclone—an area of high pressure marked by closed isobars with clockwise motion of the air.

"Most meteorologists are agreed that a circumscribed area of barometric depression is usually a locus of light ascending currents, and therefore of an indraught of surface-winds which create a retrograde whirl (in our hemisphere)....Conversely we ought to admit that a similar area of barometric elevation is usually a locus of dense descending currents, and therefore of a dispersion of a cold dry atmosphere, plunging from the higher regions upon the surface of the earth, which, flowing away radially on all sides, becomes at length imbued with a lateral motion due to the above-mentioned cause, though acting in a different manner and in opposite directions."

Sir Francis Galton, 1863

Ridge of high pressure or **wedge**—a local extension into U or V form of the isobars of an area of high pressure.

Col—the region between two cyclones alternating with two anticyclones where the isobars present a bulging surface towards a central point like the lines of a family of curves known as hyperbolas.

CHAPTER V

THE WEATHER-MAP PRESENTS THE PLAY

Science moves, but slowly, slowly.

TENNYSON, *Locksley Hall*

I. BAROMETER-SOLO ACCOMPANIED BY WIND AND WEATHER

The material to be presented

So far we have given the reader some information about the ways and means of the principal actors of the weather-drama and have indicated the vague possibility of expecting the future repetition of their behaviour with its inherent difficulties; and now we must realise that on the atmospheric stage, as in the artist's drama, the play itself is the thing, and look at the way in which meteorologists endeavour to visualise it.

Obviously we must bring on to the stage all the actors whose turn it is to take part and, as all the world is our stage, our ideal must be to develop an instantaneous picture of what is going on in weather all over the world. For that we may imagine the members of its vast auditory who are in the official reporters' gallery, and some in private boxes, to make simultaneous notes of what is visible from their points of view, and put it all down on a map of the stage, the world. In ordinary prose, at the same instant of time on land or sea, observers in every part of the world should go to their instruments with their note-books, make their observations, code them for transmission and pass them on by wire or wireless to the nearest central office for retransmission to all the establishments that want a complete map—a magnificent feat of international co-operation. But always one-third of the world is asleep if the other two-thirds are awake; and except on two days in every year, a large quarter of it has more daylight than darkness and another large quarter, more darkness than daylight. Between these opposite polar quarters is the intertropical belt where summer and winter have lost a good deal of their meaning and where when there is any display of emotion it takes, for the most part, the tempestuous form of a thunder-storm or a hurricane.

There is a limit to what can be done even by international enterprise, gradually extending it is true but still somewhat restricted. For simplicity's sake let us have permission to leave the weather of the inter-tropical zone out of our map and hope we do not thereby lose the key to the behaviour of our own weather, though we shall ascribe the striking features of its activity to the interplay of the air that comes from the intertropical zone with that which comes from the opposite quarter—the frozen north (Chap. VI).

For some reason or other, countries like to have their waking, or their working day, begun with an observation of the weather and closed with another, and they are disposed to pay less attention to what happens in the intervals; so what is convenient for places which keep Greenwich time is not really convenient for the western hemisphere that has a standard time five hours later than that of Greenwich, nor for countries east of British India which have had six hours of their day before London is officially awake. So, in our time, strictly simultaneous maps of weather are not exactly maps of the world. Still it was a great achievement to get simultaneous observations for the whole of Europe and North Africa at 7 o'clock of the Greenwich morning, 8 o'clock of the Scandinavian, German and Italian day and 9 o'clock of the eastern European and Egyptian day. And now a map for 1 a.m. occupies the place of honour.

And a similar achievement at 8 a.m. and 8 p.m. of 75th meridian time for the United States, Canada, the West Indies and Mexico, with ships at sea assisting both sides in an organised plan, is indeed something that the man-in-the-street may, if he will, put into comparison with the impressive list of the movements of liners, the casualties to ships or the despatches of mails, which escape his ordinary notice in the small print of his daily paper, or the prices of stocks and shares which have indeed a wider appeal even than the Daily Weather Report.

The play of wind and weather goes on without any intervals. The shades of night have indeed some effect upon the weather but they do not stop the progress of events. When the sun is rising in our east it is setting in somebody else's west, and the perfect map would picture the action of the drama throughout the 24 hours; but already the London Meteorological Office devotes ten large pages to the weather of 24 hours, and that limitation is only achieved by selecting 1h, 7h, 13h, 18h, four of the 24 hours of Greenwich time as observation hours. If it were possible or sufficiently convenient to observe at the hours of 6h, 12h, 18h, 24h of Greenwich time, all over the world, all parts of the world might take a nearly equal share of the inevitable inconvenience. Ships at sea which are always on the watch could join with observers on land to produce a plausible story of the progress of the play.

Already the ships are in line, Europe is an hour late at three of the observation times, the observers of the United States and Canada are at their posts at 1h and 13h, though that may mean 3 p.m. and 3 a.m. in Alaska; stations are being established beyond the Arctic circle, so gradually the ideal is being approached. In 1937 the North Pole was included in the map as now shown in our frontispiece.

The mode of presentation

For each observer the turn of the hour means an observation of pressure, wind, weather, temperature, humidity, visibility, state of the sky as regards cloud, state of the ground or sea, with notes of past weather, maximum and minimum temperature, rainfall and at some stations sunshine. We have already set out a specimen of the day's observations on p. 132. As many as possible of these have to be got on to the map. The work has to be very promptly and expeditiously done. In London the 7 o'clock map of Europe, the Atlantic and North Africa is ready between 8 and 9 o'clock. With observations from Asiatic Russia on the one side and from America on the other, aided by incidental observations of the pressure, temperature and wind in the upper air from air-planes or balloons in various parts of Europe; forecasts are issued by 10 o'clock, the day's report completed by about 12 o'clock, printed and in the post by 1 o'clock, reaching recipients within short range at about 4 o'clock. The result is a printed map of nearly the whole of the northern hemisphere within about 5 hours of the latest observations. It is a specimen of this map for 25 February 1933 (fig. 81) which we suggest to the reader as typical of how the meteorologist presents the play. The pressure, measured in millibars and "reduced to sea-level" by a formula, is indicated by lines of

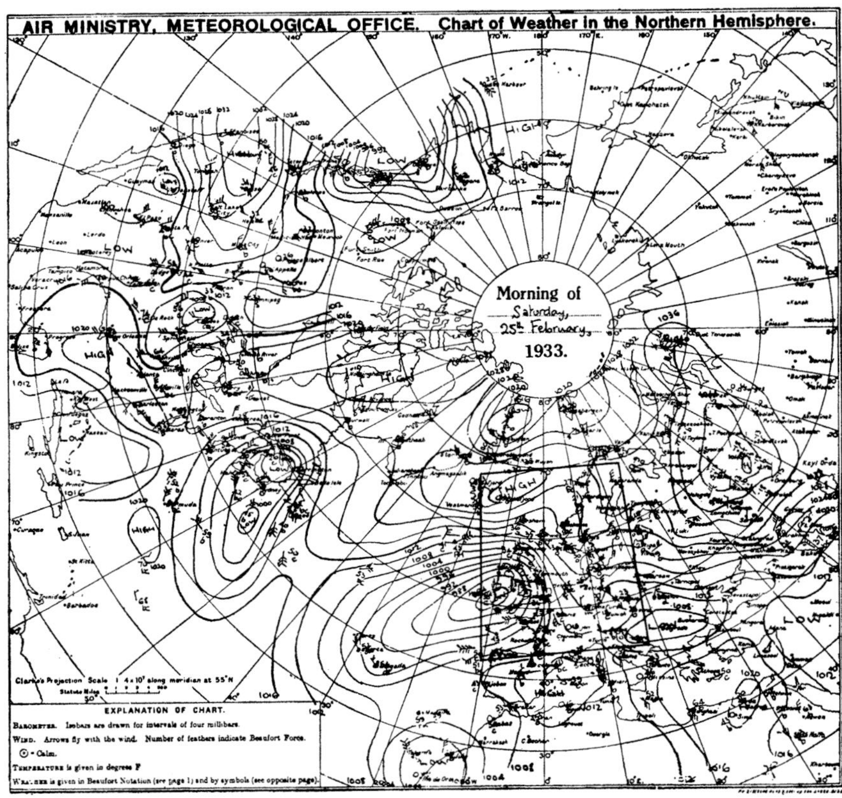

Fig. 81. Map of the weather over the greater part of the Northern hemisphere for 25 February 1933 representing conditions at 7 a.m. Greenwich Mean Time over Europe and North Africa and 1 a.m. G.M.T. over North America. From the report issued daily by the Meteorological Office, Air Ministry, London, price 1*d*.

equal pressure—sea-level isobars—which sometimes cross mountains as if they were aeroplanes, the direction of wind at a station by an arrow, the force being indicated by the number of feathers, temperature is given in figures of the Fahrenheit scale, weather by a symbol at the station or a letter underneath the temperature figure. Humidity does

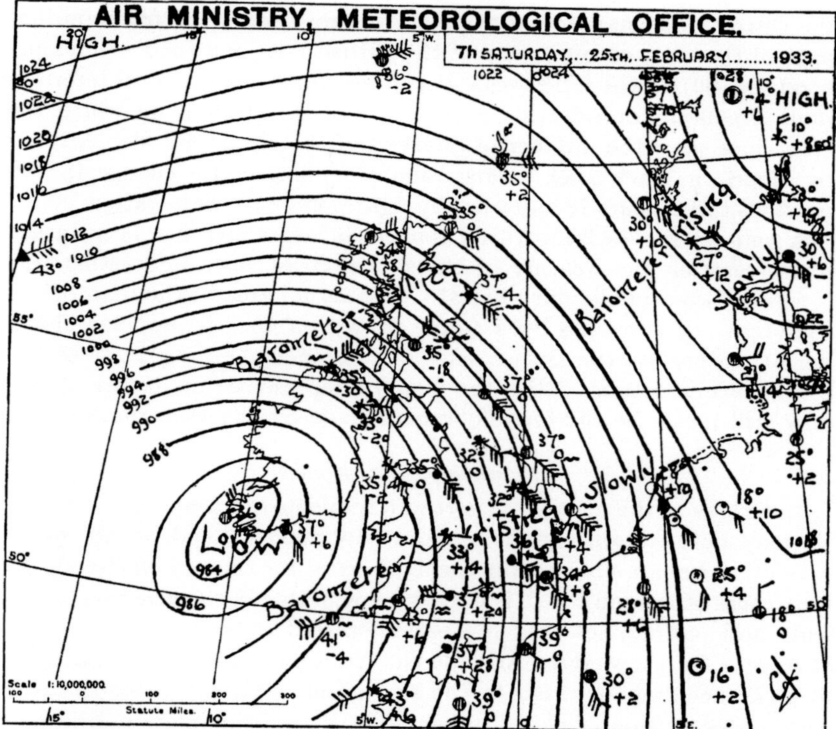

Fig. 82. Detailed weather-map of the region of the British Isles from the *Daily Weather Report* of the Meteorological Office, Air Ministry, 25 February 1933 7 a.m. G.M.T.

The announcement of the thaw that converted the snow of the 24th into the floods of the 26th.

The frame drawn on the circumpolar chart opposite, to mark the area covered by the map of the Victorian period, extends farther north, east and south than this but not so far west.

not appear in the display. Perhaps it ought to; but at any rate the forecaster is aware of it. Perhaps, again, dew-point is the best indication.

The reduction of scale and the confusion of the colours of the original lithograph in the photographic reproduction make it difficult to recognise the symbolism of local details of the circumpolar chart. Accordingly for the clearer understanding of the plan we are reproducing from another page of the same report a map of the British Isles and their neighbourhood (fig. 82) on the larger scale issued for the display of local information. Small circles indicate the positions of the stations on which the cross-lines show the number of quarters of the observer's sky covered by cloud; a black spot means rain. The barometric change in the previous three hours is indicated by figures close to each station.

A map on similar lines with a somewhat larger area has been in use for forecasting the weather of the British Isles since forecasting was recognised as a public duty in 1879. The actual area of the map of the Victorian era is shown by a frame drawn on the circumpolar chart, fig. 81, to enclose the British Isles and the neighbouring parts of the continent and oceans. With the modern representation before us we may still recognise the forcible appeal made by the Victorian map.

Cyclones and anticyclones

If we look at the map of the northern hemisphere represented in fig. 81 we may note certain regions marked LOW and enclosed by an isobar with other isobars surrounding them. There is one centred on the south-west coast of Ireland which is clearly recognisable in the chart on larger scale, and another, not very impressive, centred on the north-west coast of Africa. On the western Atlantic there is a pair, one over the Gulf of St Lawrence and the other to the south-east of it, the two contained within a succession of closed isobars. There are incomplete indications of another off the coast of Alaska. As it happens those are the fashionable places for growing what the Victorians called cyclones.

Thinking now of the one which was affecting Britain and which carries a figure of 984 mb, and looking up towards the north, a little searching will identify an isobar of 1016 mb centred on the east coast of Greenland, not exactly low but lower than its next neighbour; and between that low and the one centred on the Irish coast is another closed isobar which is carrying the label HIGH and is marked 1032 mb. That also is surrounded by an enclosing isobar, not completed on the map. And if we look further east we shall find another enclosing set round a closed isobar marked H with 1048 mb discernible with a magnifying glass.

Thus there are centres of low pressure and centres of high pressure surrounded by closed isobars which make easily recognisable features on the map as cyclones or cyclonic depressions and anticyclones.

If we could examine the winds that are associated with the closed isobars we should find that it is not unfair to say that the general movement of the winds in the region of low pressure is "counter-clockwise", that is round the centre in the direction opposite to the way clock-hands move. And on the other hand the winds associated with the closed isobars round high pressure move clockwise and go round the centre in the same direction as clock-hands.

These two types of isobaric distribution easily attract attention when they appear on the map. They take the leading parts on the weather-stage while the supplementary areas that help to fill the map get less attention. The isobars lying between the high and the low, that may run a course as straight isobars for hundreds of miles, represent in fact the greater part of the energy of the picture but they are hardly noticeable in the general map of the hemisphere. They found recognition more easily on the more limited geographical area and larger scale of the Victorian chart.

On the map of the hemisphere one may study with advantage the region known as "col" where the curving isobars of two lows face one another with a corresponding pair of curving isobars of two areas of high pressure also facing one another on the cross-line. There is an example shown in the middle of the North Atlantic Ocean in fig. 81 and another, often a very important one for British weather, over the ocean outside the Strait of Gibraltar between Madeira and the Azores. The situation is easily described by the geographical analogy from which the name col is borrowed, a pass providing easy access from one valley to another between two hills one on either side. The situation shows a small nearly level area higher than the valleys which it separates and lower than the ridges on either side. So at the col pressure is nearly uniform over a small area, higher than in the lows between which it lies and lower than the highs on either side; and so far as air-currents are concerned two lead towards the col and two away from it and the possible varieties in the nature and behaviour of the airs meeting on the col may lead to a good deal of meteorological excitement. It is more versatile than the more pompous depression. It has been appealed to in explanation of November fogs.

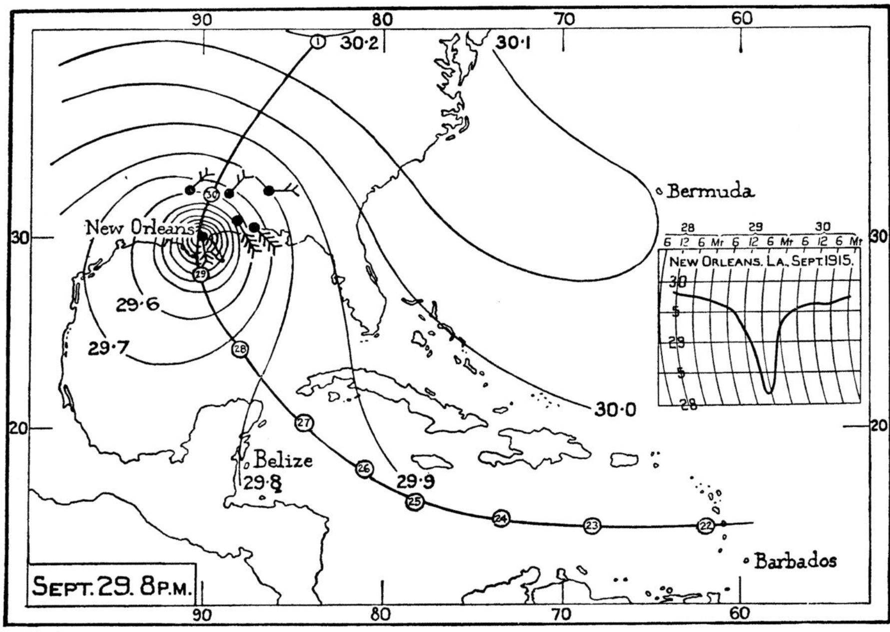

Fig. 83. Chart of a tropical cyclone travelling from Barbados to New Orleans in eight days, and passing on across the United States, with an inset showing the change of the barometer during two days when the cyclone was near New Orleans. The numbers in the small circles mark the position of the centre on the corresponding days of the month.

Of the component parts, the protuberances of high pressure towards the col are known as ridges or wedges and the protuberances of low pressure lying between them are troughs or V-shaped depressions.

These all took their part in the version of the drama that held the stage in the last quarter of the nineteenth century and were regarded as associated with special types of weather. But the closed isobars of the cyclone and anticyclone were the star parts of the period.

Analogy with hurricanes and tornadoes

In the earlier part of the nineteenth century a great deal of attention was paid by seamen to the tropical hurricanes of the West Indies, the Indian Ocean and the Western Pacific which were extremely dangerous for shipping and indeed are still. With waterspouts and tornadoes they came to be regarded as travelling vortices of air for which physicists could find an analogy in the occasional smoke-ring from a tobacco pipe or the corresponding product of a special apparatus to construct them, or in the eddy of the street-corner or of the dusty tropical plain. And a good deal of the energy of mathematical physicists was devoted to the study of vortex-motion; even atoms came to be regarded as amenable to a vortex-theory.

In those days it was not easy to get a travelling hurricane mapped on a weather-chart but the ideas of its constitution are now quite effectively represented in fig. 83 with its closed isobars surrounding a conspicuous "low", and its daily travel.

Those of the Indian seas were called cyclones and carried the suggestion of spiral motion towards the centre. The Victorian meteorologists brought the name into use for the groups of closed isobars round centres of low pressure on their own maps and coined the name anticyclone for the closed isobars round centres of high pressure.

This episode of the study of the drama began with Le Verrier's investigation of the travel of a storm that damaged the allied fleets at Sebastopol in 1854 and was taken up by FitzRoy in the investigation of the storm in which the *Royal Charter* was lost in 1859. From 1867 to 1879 after the daily weather-map had been introduced every so-called cyclone and anticyclone that crossed the British Isles was watched—set in a note-book, learned and conned by rote—analysed by the autographic records of instruments set up for that very purpose.

The study of weather-maps soon disclosed the fact that the cyclones shown on them also travelled. But with the tropical hurricane the travel of the centre, about ten miles an hour, was slow compared with the speed of the wind that might reach 100 miles an hour, whereas the travel of the cyclonic depression, north of 40° N, was generally as fast as that of its fastest wind.

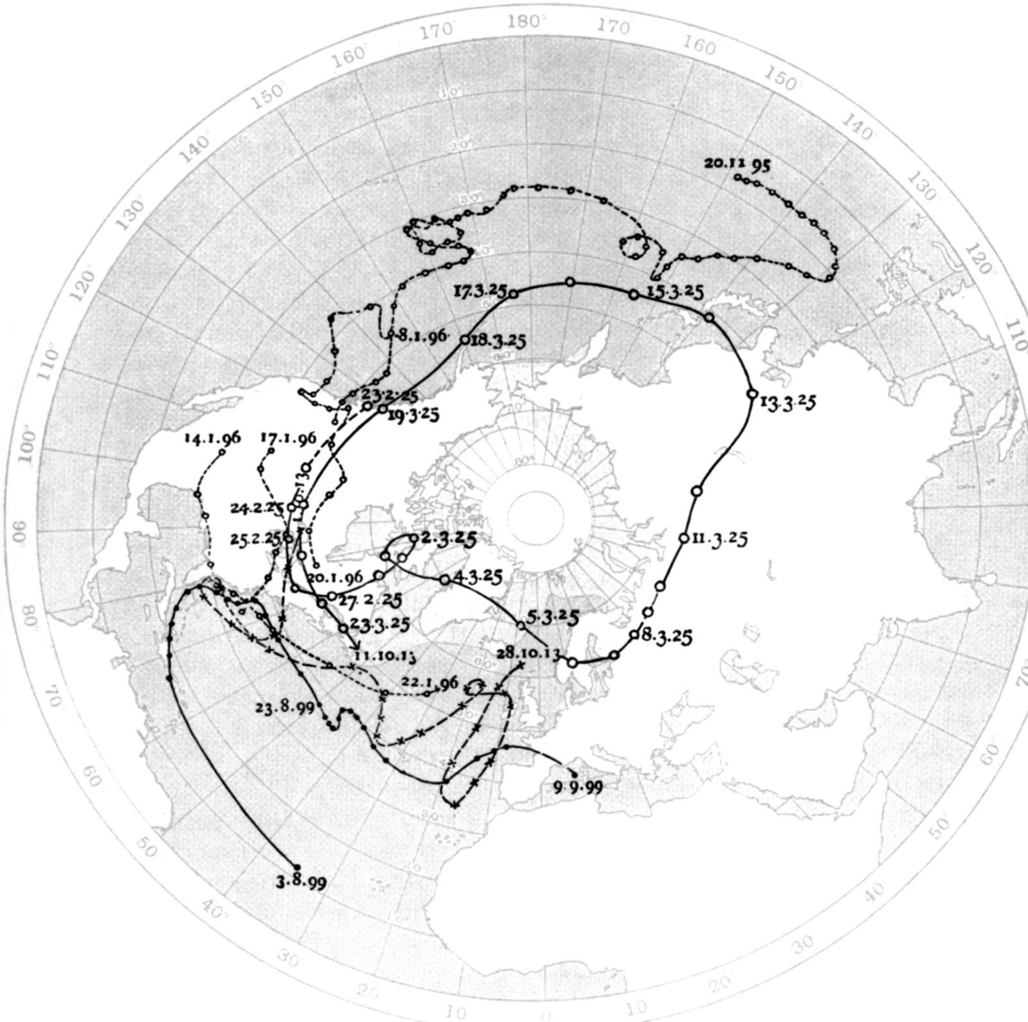

Fig. 84. Paths of cyclonic depressions with long life-histories.

1. From the Western Pacific on 20.11.95 to middle North Atlantic on 22.1.96, with duplication in the Pacific after 8.1.96 and redistribution on 14.1.96 at the Rocky Mountain barrier (McAdie).

2. From the Great Lakes 1.10.13 by an irregular path to the Faroë on 28.10.13 (McAdie).

3. From the equatorial Atlantic, long. 35° W, on 3.8.99 to W. Indies, long. 67° W, on 8.8.99 and by an ordinary path of hurricanes and depressions to the Mediterranean on 9.9.99 (M.O.).

4. From Vancouver Island 23.2.25 and Canadian border 24.2.25 round the world to Strait of Belle Isle on 23.3.25 (Mitchell).

Each of the four paths has its line specially marked and dated at intervals.

The travel of centres with their weather

The general result of the investigation amounted to accepting that weather travels with the centres and that the travel can be expressed by the successive positions of the centres of cyclonic depressions, or, with less assurance, of those of high pressure.

Fig. 84 shows some examples of the tracks of depressions that have been derived from the examination of successive maps either in England or in America. It is important to notice that the track from 3.8.99 to 9.9.99 began as a tropical hurricane and became a more temperate vortex after it had rounded the western end of the permanent Atlantic high pressure and had taken up travel eastward as an ordinary Atlantic depression.

The one which started from the western Pacific 20.11.95 and finished up in the Atlantic Ocean on 22.1.96 seems likewise to have changed its type. However hazardous the identification may seem to modern eyes the idea then was that we had only to understand what weather was appropriate to the different groups of isobars and the rest would follow. Scientific forecasting was possible.

The barometer and its isobars

Let us examine the matter a bit. On a barograph the passage of the cyclonic phenomena is indicated by a fall of pressure succeeded by a recovery which generally takes about a day to accomplish and so a "depression" is shown on the record, and the name depression became associated with the condition under which the local load of the earth's atmosphere was really *lightened* by the removal of something between less than 1 per cent and 10 per cent of the normal load; the change of pressure may be between 10 mb or less, and 100 mb.[1] In more modern days the name depression is taking the place of cyclone for areas marked by closed isobars with counter-clockwise winds surrounding centres of low pressure in the region of the earth where these depressions are formed by air from the tropics coming into conjunction with air from the polar regions.

[1] The amount of air removed to produce depressions has been estimated in millions of tons as 40,000 for the hurricane of p. 199; 6600 for that of p. 224; 86,000 for a North Sea depression in 1917; 2,100,000 for an Atlantic depression in 1926.

VICTORIAN BAROMETRIC SOLO

Arranged by R. Abercromby and W. Marriott.

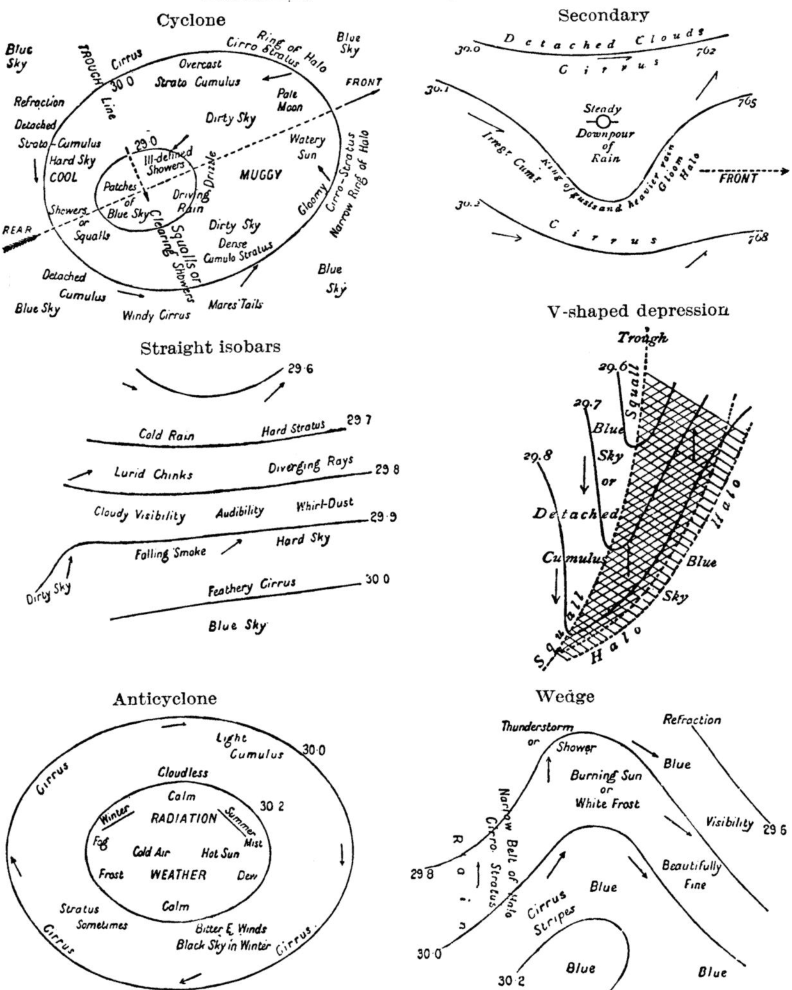

Fig. 85. Prognostics based upon barometric distribution (1882–7). The diagram of the cyclone has been modified by the restriction of the trough-line to the right-hand side of the line of travel. In the diagram of the V-shaped depression cross-lines indicate cloud and double cross-lines rain.

It was recognised that the winds associated with local sinuosities, mere bends in the isobars (very suggestive of depression in the barograph trace), carry weather with them which has the same characteristics as that of closed isobars and so are described as secondary depressions. The whole drama of the weather comes to be associated with the ground-plan of the isobars shown on the map, and expressed by the movement of cyclonic depressions, secondaries and anticyclones across the map.

So the practice arose of regarding the weather of the epoch as associated with the forms of isobars which can be recognised on the map, and it is not unfair to describe the Victorian meteorologist's idea of the play as wind and weather accompanying a barometer-solo.

The solo

The attendance of wind and weather upon the vagaries of the barometer was set out by various authors and especially in a book entitled *Weather* by the Hon. Ralph Abercromby, an experienced seaman and competent meteorologist of the Victorian era. Proper names were assigned to recognisable groups of isobars which were regarded as having a definite rôle to play in the drama: anticyclone or high pressure, cyclone or cyclonic depression, secondary depression, straight isobars, trough of low pressure, wedge of high pressure, col between two highs and two lows—all are exemplified on the map of fig. 81 if the reader will look for them. The cyclonic depression and anticyclone were the chief, and have remained the principal actors on the stage between the latitudes of 35° and 60°, so that to this day the announcement of a "forecast for to-night and to-morrow" begins with the information that pressure is high over Europe and a depression off the west of Ireland is moving north-eastward, or the anticyclone lying between Britain and Iceland is stationary and a smaller anticyclone over Scandinavia is moving southward.

The idea was that as the various barometric distributions moved they carried weather with them. Abercromby's indications of the weather to be expected in association with the different types of isobaric distribution are given in fig. 85 except that there is no representation of a col. But the materials of which a col is composed are shown as V-shaped depression and wedge. We may supplement his anticipations by the

Fig. 86. Some celebrities of the barometer-solo.

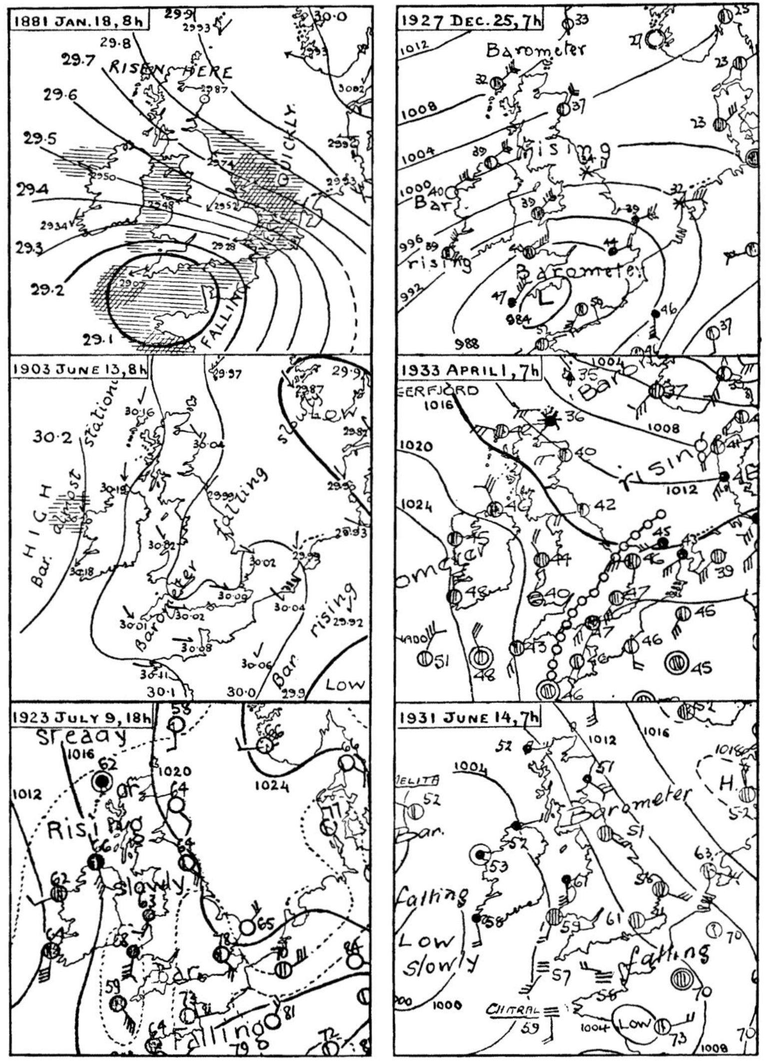

Some other examples, with estimates of energy, are given in Chapter VI.

representation in fig. 86 of a few striking passages of the solo as represented on the maps of the past 50 years.

Celebrities of the barometer-solo

1. **18 January 1881.** The isobars of a depression which passed up the English Channel and paralysed traffic in the south of England for three days. In a recent note its reputation is said to have been exaggerated because it happened to London, but an 'eye-witness' has explained that, invited to a dance that evening at Uxbridge, on the journey from Cambridge in the morning he found the snow-crystals drifting through the chinks of the door and window of the railway coach, and the platform of King's Cross, underground, covered with snow. Arrived at West Drayton in the afternoon the only available traffic to Uxbridge was on foot "by the field path", rail and road both impassable. Spending the night at the inn at West Drayton, the 19th took him to Reading where another night was spent in the company of the engineer in charge of getting a cattle train out of the snow at Goring. At 5 p.m. the next day the engineer authorised a ticket as far as Oxford—a wonderful sight in the snow—and from there northward only light snow was found; but four days from London to Birmingham is fortunately exceptional.

2. **25 December 1927.** A similar occurrence that made things difficult on Christmas night in London and during the following week. Kentish villages were supplied with food by aircraft. The statistics of gales in the Meteorological Office suggest that the weather is apt to 'break' in that way between Christmas and the New Year.

3. **13 June 1903.** The precursor of three days' continuous rainfall in London. So persistent was the fall that the forecaster on duty felt a good deal of hesitation about repeating the interpretation of the signs.

4. **1 April 1933.** The precursor of wonderful weather, warm sunny days of middle spring which by the end of the month had brought out the chestnut blooms about a month before the usual date. As a map the example is noteworthy because it carries a new symbol, small circles linked together by lines, a new histrionic feature of the

weather-map introduced on 1 March of that year with the name "occlusion front" to which a more formal introduction will be given to the reader on p. 253.

5. **9 July 1923.** The precursor of very violent and persistent thunderstorms in the night possibly taking time to await a supply of water from the south coast.

6. **14 June 1931.** The precursor of a "whirlwind" destructive enough to be called a tornado at Erdington near Birmingham.

Very small "depressions" with remarkably violent winds, the British equivalent of tornado, are formed sometimes as secondaries within the isobars of a larger depression and travelling along the isobars of the main depression cause a good deal of damage to trees and buildings. One passed over Cambridge from Southern Ireland WNW to North Norfolk on 24 March 1895, another from Devonshire northwards along the Welsh border on 27 October 1913. To obtain the necessary pressure difference the local reduction of pressure must extend to very great heights.

II. THE DISAPPOINTMENT OF THE NINETIES

Abercromby's book is dated 1887 and from the rules and examples which he set out it might appear that, with the recognition of the rôle of cyclone and anticyclone, secondaries and other varieties of depression, the problem of to-morrow's weather was solved. Anyone who had a weather-map might be his own forecaster. The enthusiasm of the eighties following the twelve years of careful investigation was quite remarkable. Storms travelling across the Atlantic might be announced from New York perhaps four days ahead. The London *Times* and the *New York Herald* joined in an endeavour to realise the project and so anticipate the weather of our western seaboard which marked the western limit of our own information.

But the result was in the end disappointing; it might be true that depressions travelled but they changed on the way without any apparent rule or reason. Daily weather-maps of the Atlantic compiled in the Meteorological Office for the Polar Year 1882–3 made their behaviour clear but incalculable. Before the end of the century British enthusiasm had evaporated, and the issue of maps and forecasts became a matter of routine. So far as one can judge it was the same in France. The subject was still actively pursued in the United States, in Germany, in Norway, Sweden and Denmark, less actively in Austria, Italy, Switzerland and Spain. Austria and Russia were perhaps more interested in climate than in weather and in Britain it was nearly agreed that marine charts were the really meritorious service of the Weather Office.

It may be of interest to note some points in the Victorian practice that have proved to be snags in the course of science.

The flow of air

One condition of misfit may be attributed to the idea that the flow of air on the map would naturally take the form of a current from high pressure to low. The idea easily arises from the conception that the plot of the drama of weather begins with the local reduction of pressure by the ascent of warm moist air to form the central region of a depression and the inward flow from the higher pressure of the environment to keep the process in operation.

Van Bebber, 1890.

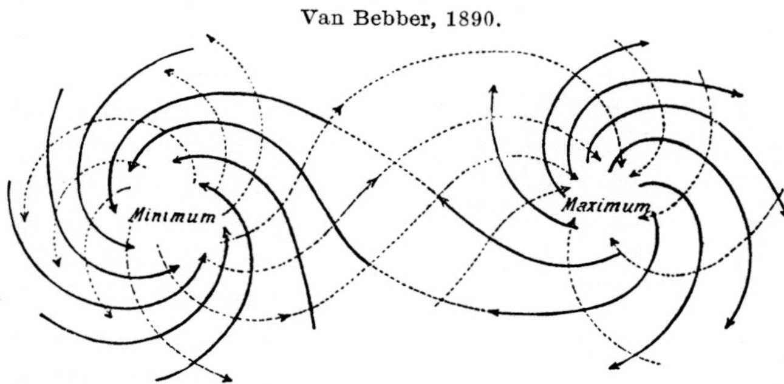

Fig. 87. Diagram of the circulation of air along spiral curves between high and low pressure. The full lines represent the motion at the surface, the dotted lines the motion of the upper air.

That may seem quite reasonable if we turn our attention to the direction of the arrows which represent the winds. They suggest a spiral path from a centre of high pressure to the centre of the low. Towards the end of the century this idea was elaborated by Dr W. J.

van Bebber of the Marine Observatory, Hamburg, with a picture of the circulation of air between high and low pressure, along spiral paths from high to low at the surface and from low to high above. The picture is reproduced in fig. 87.

But the idea which the map so easily suggests is really an illusion. If the scene which our map represents were permanent and everything were the same to-morrow and the day after as to-day except that the air had stepped along the lines drawn for its paths, the moving air would not only appear to seek the centres of low pressure but would actually find them. But as we take the successive maps the centres of high pressure and low pressure will be found to have changed their positions and what looks like a good aim to-day will have proved to-morrow to be a bad miss.

With the turn of the present century the actual motion of the air in the neighbourhood of cyclonic and anticyclonic centres came to be examined and tracks set out; the idea that was evolved is represented by fig. 88 and from that a number of inferences could be drawn.

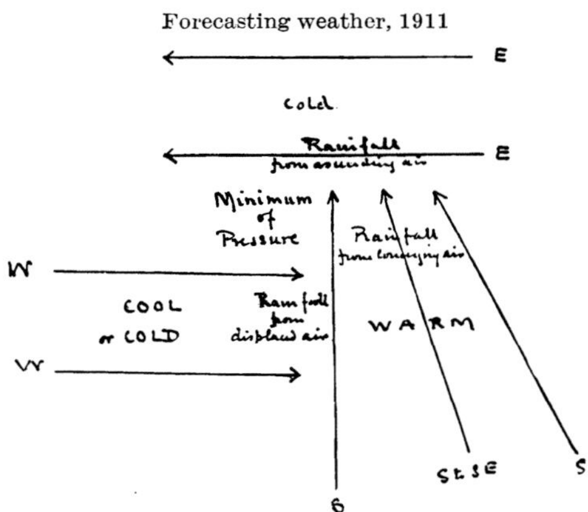

Fig. 88. Original sketch for the distribution of wind, temperature and rain with reference to the centre of a cyclonic depression.

First in a well-developed depression the winds do not circulate round the centre, the warm south wind seems to be making direct for the flank of a colder easterly current that is crossing its path and, in its turn, to be taken in the flank by a colder wind shown as coming from the west but possibly round from north-west or north and exhibiting the features of Abercromby's trough on the line drawn southward from the centre but without any corresponding experience along the line drawn northward. The same kind of experience is exhibited in the line of the V-shaped depression.

The smooth run of the nicely rounded lines of the isobars in a Victorian map might also be illusory as the map-maker had only observations of pressure at stations perhaps a hundred miles apart to guide his pencil. Intruding changes are of small magnitude; they might easily be and indeed were overlooked.

The rule of the road in the free air

To a certain extent the practice of drawing isobars at sea-level over the land encouraged laxity about the run of the lines.

In the early days of weather-maps C. H. D. Buys Ballot, Professor of Physics at Utrecht in Holland where there is not much difference of level, pointed out that the motion of air was more reasonably regarded as a flow *along the isobars* keeping the lower pressure on the left. The line of the wind showed a deviation of 60° or 80° from the line to the centre of low pressure.

This was indeed a most important contribution to the dynamics of the atmosphere. It is known as Buys Ballot's law and leads to a better conception of the rule of the road for air. It has indeed led to the suggestion that in the free air there may be a definite relation between the flow of air along isobars and the distribution of pressure, a flow which at the surface would be disturbed and impaired by the obstacles that have to be traversed.

In Victorian days it was the behaviour of the centre of lowest pressure or of highest pressure which was used to identify the action of the drama, and as a matter of fact they are just those parts of the map where the motion of the atmosphere is least noticeable. On our ordinary conception of motion we might think of them as the parts which must be left behind because of their inactivity. For the Victorian meteorologist, however, thinking of tornadoes and hurricanes, the centres were "eyes" of the storm and it was the centre of the cyclone or cyclonic depression that "rides in the whirlwind and directs the storm" and not the vagrant winds that were merely keeping an eye on affairs of the ground. In those days convection might be trusted to provide the distribution of pressure and the winds to adjust themselves thereto.

In the play itself there are changes going on in everything, in the pressure, in the wind, in temperature, in humidity, in the state of the sky; but it is only the winds that would impress the observer with the idea of motion, generally carrying clouds along with them. We know that if pressure changes, it must be because the air, of which the weight is the local pressure, has moved. Some has come from or gone to another place carried somehow by wind; temperature and humidity may change on the spot without motion. But the motion of the wind is very real. To stop the motion of a strip of air between two isobars, the width 1 mile (less than ·001 inch on the map) and thickness on the same scale not greater than that of the ink-mark would require a barricade strong enough to stop 400 five hundred ton trains running at 50 miles an hour. Considering the enormous number of such strips that would make up the whole map the driving-power of the motion represented is quite stupendous.

It is these enormous moving forces of the wind that raise waves, drive ships, carry clouds and keep the weather of the world in being. And what we see pictured on the map as wind is only what remains at the surface when the obstacles there have done what in them lies to reduce the motion. Half a kilometre above the surface of the sea the winds of the free air can be estimated as having roughly twice the speed of that which the ship feels, and over the land three times the motion that operates a windmill. The models of fig. 52 are eloquent on this point.

Perhaps we ought rather to think of the action of the play as the continuous formation of a new centre by the vagrant winds, the real actors moving across the stage, than as the behaviour of the most inactive of the whole company.

The idea gains a good deal of strength from a consideration of the rule of the road for currents in the free air. Let us think of a pair of air-currents passing each other, on their opposite ways, in relation to the paradoxical rule for British road-traffic: "If you go right you go wrong and if you go left you go right". If the air currents keep this rule and, going left, leave each other on the right the weather between them will be anticyclonic—serene, calm and sunny or perhaps foggy, warm in summer often cold in winter[1]. But if they follow the continental rule, go right and leave each other on the left as they pass, cyclonic weather will develop between them and with the water-power in reserve there is no limit to the restless energy that may be displayed by the wind. This is the more reasonable if we remember that for a warm west wind to pass a cold east wind by keeping to the left it must get to the north of its opposite; if it keeps south of it there is trouble.

If we might go a step further we may have a triangle of straight isobars the winds of which in the free air maintain the varying fortunes of a cyclonic depression between them. Such a depression might respond to variations, like our familiar friend the "depression over the North Atlantic". This idea is developed in Chap. VI.

Some new ideas

With the turn of the century also there came from Sweden the suggestion that the alterations in the pressure within the past few hours gave a useful indication of coming changes in the map and even deserved a separate map for themselves. So the tables of observations from which the map of fig. 81 is compiled have a column for the change of pressure at each station within the three hours immediately preceding the observation. The changes are marked on the map of to-day by an entry close to the position of the station.

And with the insistent requirements of aircraft for information about the state of the atmosphere as regards the visibility of distant objects, visibility also has now a column of its own in every table of observations and is regularly included in the forecasts for shipping.

With the turn of the century too, the area covered by the map began to widen. First came observations from the Azores, the Portuguese islands in the middle of the North Atlantic, then from Madeira, and then from the Danish islands Iceland and the Faroë. Then came messages by wireless from ships at sea and then from America and Greenland until now very little of the Northern hemisphere is left unrepresented.

[1] A remarkable example of the track of an air-current 3500 miles long from Bermuda to Norway with an opposite current on its right and sunny weather over Britain is shown in the maps for the first week in June 1933.

The subject is approached to-day with wider views and more experience than were possible with the Victorian map. So let us turn our attention to the maps of to-day in order to obtain a more modern view of the relation of the weather of yesterday with the weather of to-morrow. We will use the maps for 23, 24 and 25 February 1933 as examples.

When we were thinking about the pageantry of the sky in the Prologue we suggested that if an observer watching the clouds formed an impression of the scene and then closed his eyes for a minute he might not recognise the scene on opening them. Now let us think of the interval between yesterday's map with its highs and lows and to-day's map as an experiment of the same kind but with a very much longer interval, a day instead of a minute. And yet there is generally no difficulty in recognising in to-day's map the features of yesterday's, changed in position and somewhat altered in intensity, and even perhaps in character, but still recognisable. The depression which was yesterday between Iceland and Scotland may be to-day between Jan Mayen and Norway; but it is still alive. The main depression may be deepening or filling up, but it will almost certainly be on the map, perhaps with a new assortment of secondaries; it is still there, easily recognised in spite of minor changes. An experienced forecaster may have on his file two successive maps which are conspicuous for their differences rather than their similarities. The maps of 23, 24 February 1933 (pp. 240–1) may be cited, and curiously enough those for the same dates in 1932 show the same discordance; but such occasions are rare.

The idea of continued existence still animates the practice of forecasting the weather of to-morrow from the map of the weather of to-day in many parts of the world. Look at the actors on the stage to-day, see what their behaviour has been since yesterday and carry on to what it will produce by to-morrow. Our monthly reports still show the tracks of depressions that have travelled over our area within the month and the corresponding reports for the United States of Brazil show the travel of anticyclones in the sub-tropical belt over that country.

In acknowledgment of this practice we may still make out the habits of our star actors; for depressions, what paths they take, how fast they travel and what happens to the localities in the path of the centre. For anticyclones, more dubiously, how slowly do they move? how long may they last? Some apparently last for weeks (8–25 Feb. 1932) whereas depressions generally pass in a day.

The whole process is worked by maps and depends on maps. The Air Ministry publishes as many as six maps for each day, two for 1h and 7h and one for 13h and 18h. Without a map to guide him it is not surprising that the man-in-the-street is apt to take any statement about future weather as an oracle rather than an inference. It was on that account that in 1867 the office newly formed by government for the study of weather was called the Meteorological Office and not the Weather Office.

Unfortunately not many of the millions who listen to forecasts of weather have a map in front of them to suggest an inference instead of an oracle, but many have a barograph at hand, a weather-cock in sight and a bit of the sky to look at, and with their aid they can correlate the sequence of events with what the forecaster has predicted.

The observer's relation to the weather of yesterday and the weather of to-morrow

Let us then consider what the private observer will get on his record while the pressure-changes are travelling across the map. If the whole system were travelling we could draw a line across the map in the direction of travel and what is entered on the map along the line crossing his station would show the observer's experience. If we try it with the three maps selected for our use we shall not be surprised to learn that there are very few maps for which the changes can be indicated by motion of the whole along a single line; the motion is in practice often different for different parts of the map. Here is an example.

24 *February* 1932. The anticyclone will be stationary, the small depression over the Netherlands will move SSW, the larger depression about Jan Mayen will move SE.

So the different parts of the map are endowed with independent prospective motion and the observer must think of his position with regard to the features of the map which the forecaster has in front of

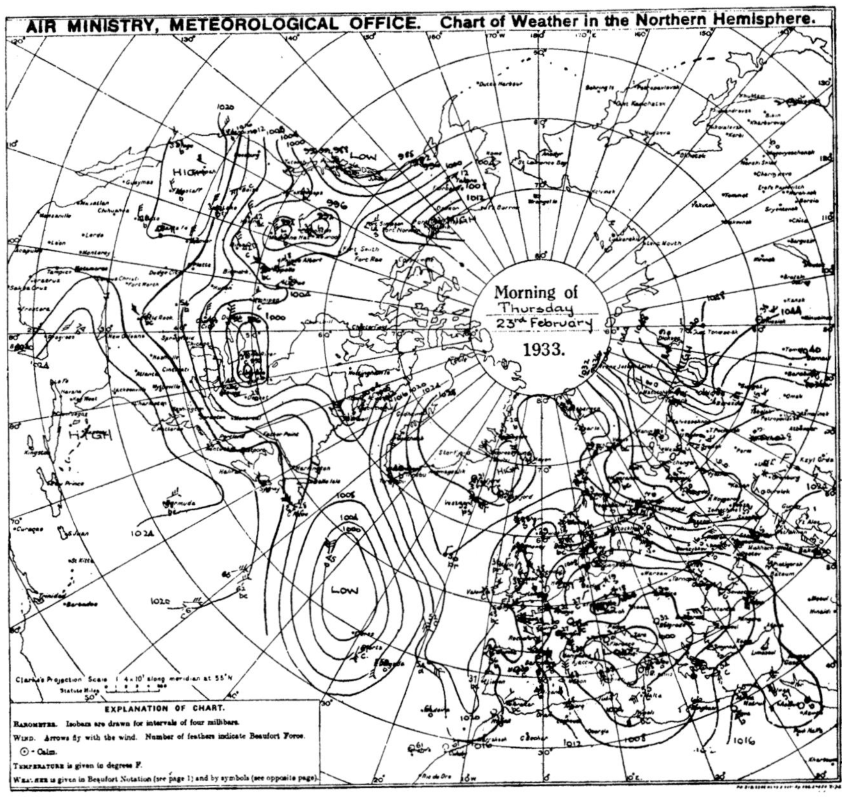

Fig. 89 *a*. Weather-map for 7h, 23 February 1933.

Readers will find it interesting to identify on these maps the persons represented according to the list with which this chapter is introduced.

him. They are indicated for the listener by the general distribution of pressure and the prospective travel of the main depression or its secondaries.

The subject is much too vast to be fully treated here. Whole files of weather-reports would be necessary for that. In practice each day has its six maps for the justification of its behaviour. But we may get some

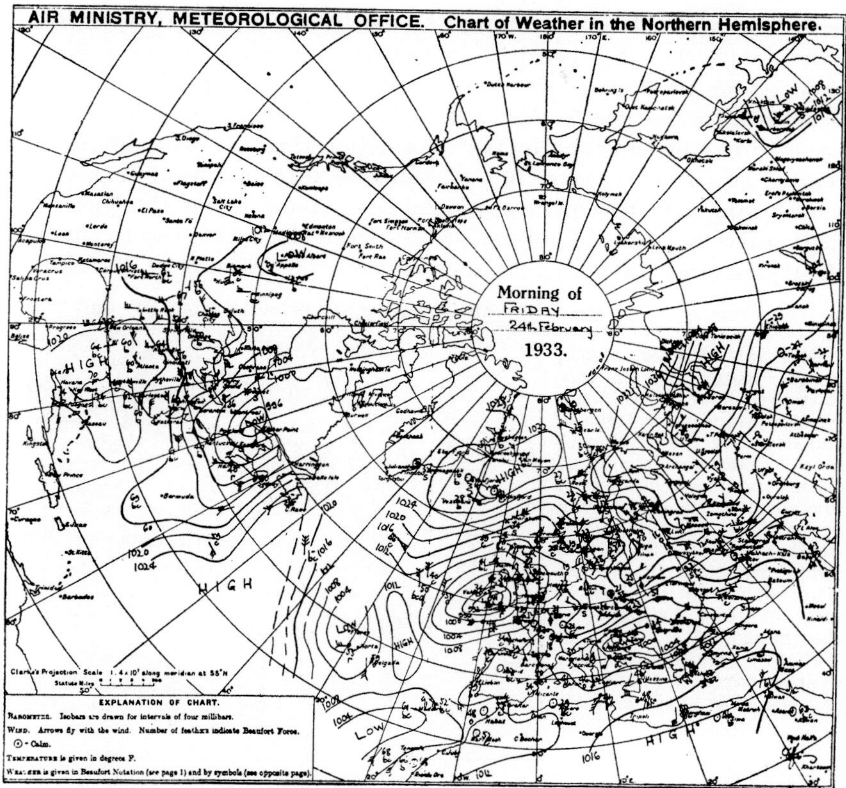

Fig. 90 *a*. Weather-map for 7h, 24 February 1933.

appreciation of the problem by an examination of a map for 7h for each of the three days in February 1933 that were associated with the appearance, arrival and conclusion of a blizzard that first covered England with snow and then deluged her with floods.

The last of the three, that for February 25, has already appeared as fig. 81. February is often an eventful month. Those for the 23rd and 24th are figures 89, 90. The episode of the blizzard is not quite complete because the low which caused the trouble is still on the stage on the 25th with its centre only a little north of where it was on the 24th; there is just enough change to suggest travelling and enough surprise at the change from the 23rd to the 24th to assure the reader that forecasting is not a mere pastime for the map-reader. The story of our three days needs to be supplemented by the information that the deep depression centred over Ireland stayed in the immediate neighbourhood more than a week, long enough to make anyone rather dubious about the travel of depressions and indeed long enough to induce the authorities to change the style of their international map as from the first day of March.

***General inference.* (23rd Feb.). Pressure is high over East Greenland, North Scandinavia and west of Ireland, and low over the Mediterranean. Comparatively small disturbances are moving west towards Britain. Winds will be mainly northerly and weather wintry. Snow will fall intermittently in eastern counties, while showers of snow or sleet are probable elsewhere.**

***Further outlook.* Wintry weather continuing, with occasional snow in the East.**

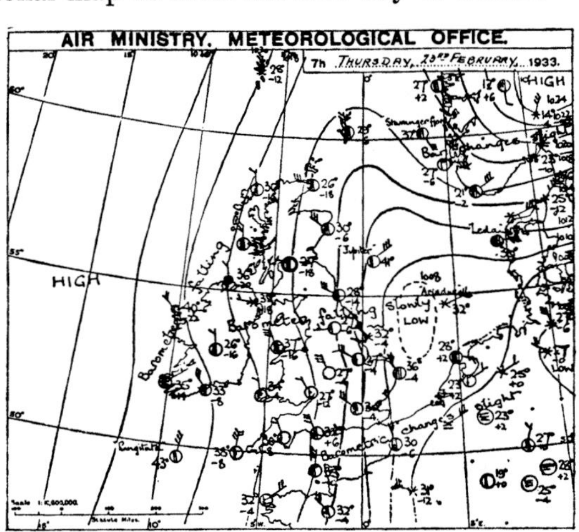

Fig. 89 *b*. Map of the British Isles accompanying the general inference and forecasts of the *Daily Weather Report* of Thursday, 23rd February.

By so doing they make acknowledgment of the practice of synoptic meteorology and the forecasting of weather which has been developed by Prof. J. Bjerknes of the Bergen Meteorological Institute with exceptional hydrodynamical aid from his father Prof. V. Bjerknes and others in the past fourteen years. Therewithal "fronts" of polar or tropical air appear in prominent positions on the stage of the weather-drama to which we will direct the reader's attention in concluding this chapter.

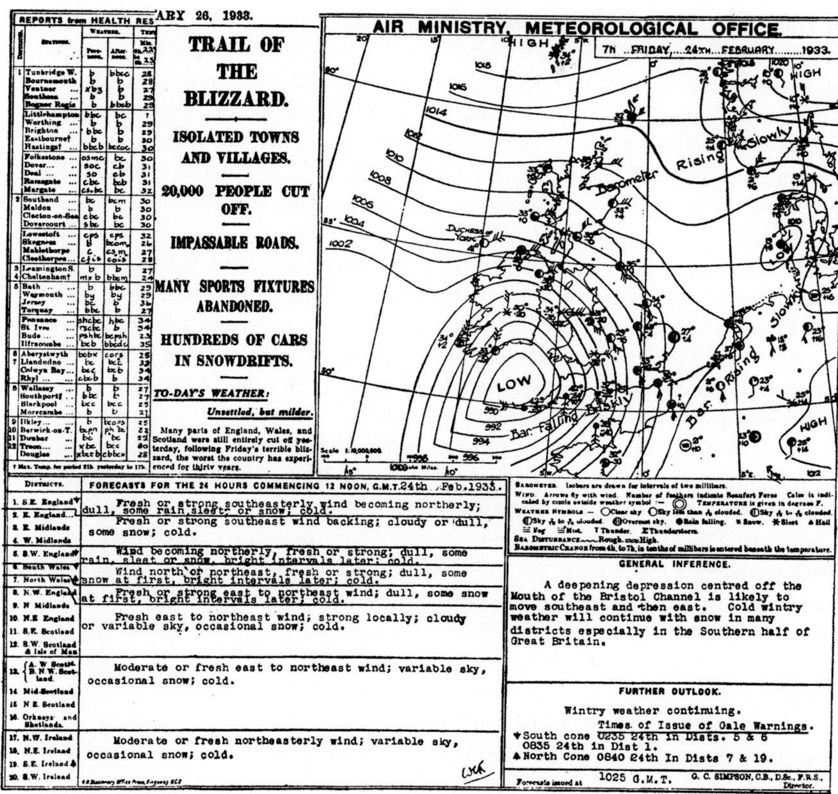

REPORTS from HEALTH RES... ARY 26, 1933.

Division.	Stations.	Weather. Fore-noon.	Weather. After-noon.	Temp. Min. 6h. 22 to 6h. 23
1	Tunbridge W.	b	bbcc	28
	Bournemouth	b	b	28
	Ventnor ...	xb3	b	27
	Southsea ...	b	b	29
	Bognor Regis	b	bbcb	29
	Littlehampton	bbc	bc	?
	Worthing ...	b	b	29
	Brighton ...	bbc	b	29
	Eastbourne†	b	b	30
	Hastings† ...	bbcb	bccoc	30
	Folkestone ...	osmc	bc	30
	Dover... ..	soc	cb	31
	Deal	so	cb	31
	Ramsgate ...	cbc	bcb	31
	Margate ...	cs.bc	bc	32
2	Southend ...	bc	bcm	30
	Maldon ...	b	b	30
	Clacton-on-Sea	cbc	bc	30
	Dovercourt ...	sbc	bc	30
	Lowestoft ...	cps	cps	32
	Skegness ...	b	bcom	26
	Mablethorpe	c	cs.m	27
	Cleethorpes...	cfis	cois	28
3	Leamington S.	b	b	27
4	Cheltenham†	mx b	bbcm	24
5	Bath	b	bbc	29
	Weymouth ...	by	by	29
	Jersey ...	bc	b	36
	Torquay ...	bbc	b	27
	Penzance ...	shcbc	hbc	34
	St. Ives ...	rscbc	b	34
	Bude	pshbc	bcpsh	23
	Ilfracombe ...	bcb	bbcdc	35
6	Aberystwyth	bcbx	cors	25
7	Llandudno ...	bc	bcc	33
	Colwyn Bay...	bcc	bcb	34
	Rhyl	cbcb	b	34
8	Wallasey ...	b	b	27
	Southport† ..	bbc	b	27
	Blackpool ...	bcc	bcc	25
	Morecambe ...	b	b	21
9	Ilkley... ...	b	bcops	25
10	Berwick-on-T.	bcpn	ph bc	22
11	Dunbar ...	bc	bc	29
12	Troon... ...	xbc	bcc	30
	Douglas ...	xbcc b	cbbcx	28

† Max. Temp. for period 21h. yesterday to 17h.

TRAIL OF THE BLIZZARD.

ISOLATED TOWNS AND VILLAGES.

20,000 PEOPLE CUT OFF.

IMPASSABLE ROADS.

MANY SPORTS FIXTURES ABANDONED.

HUNDREDS OF CARS IN SNOWDRIFTS.

TO-DAY'S WEATHER:

Unsettled, but milder.

Many parts of England, Wales, and Scotland were still entirely cut off yesterday, following Friday's terrible blizzard, the worst the country has experienced for thirty years.

AIR MINISTRY, METEOROLOGICAL OFFICE.

7h ...FRIDAY....24TH...FEBRUARY........1933.

Districts.	FORECASTS FOR THE 24 HOURS COMMENCING 12 NOON, G.M.T. 24th Feb. 1933.
1. S.E. England▼ 2. E. England	Fresh or strong southeasterly wind becoming northerly; dull, some rain, sleet, or snow; cold.
3. E. Midlands 4. W. Midlands	Fresh or strong southeast wind backing; cloudy or dull, some snow; cold.
5. S.W. England▼	Wind becoming northerly, fresh or strong; dull, some rain, sleet or snow, bright intervals later; cold.
6. South Wales▼ 7. North Wales▲	Wind north or northeast, fresh or strong; dull, some snow at first, bright intervals later; cold.
8. N.W. England	Fresh or strong east to northeast wind; dull, some snow at first, bright intervals later; cold.
9. N. Midlands 10. N.E. England 11. S.E. Scotland 12. S.W. Scotland & Isle of Man	Fresh east to northeast wind; strong locally; cloudy or variable sky, occasional snow; cold.
13. {A. W. Scotland, B. N.W. Scotland} 14. Mid-Scotland 15. N.E. Scotland 16. Orkneys and Shetlands	Moderate or fresh east to northeast wind; variable sky, occasional snow; cold.
17. N.W. Ireland 18. N.E. Ireland 19. S.E. Ireland▲ 20. S.W. Ireland	Moderate or fresh northeasterly wind; variable sky, occasional snow; cold.

BAROMETER. Isobars are drawn for intervals of two millibars.
WIND. Arrows fly with wind. Number of feathers indicate Beaufort Force. Calm is indicated by circle outside weather symbol :— ◎ TEMPERATURE is given in degrees F.
WEATHER SYMBOLS — ○ Clear sky ○ Sky less than 3/10 clouded. ◑ Sky 3/10 to 7/10 clouded. ◑ Sky 7/10 to 9/10 clouded. ● Overcast sky. ● Rain falling. * Snow. ✱ Sleet ▲ Hail ≡ Fog ≡ Mist. T Thunder. ⊼ Thunderstorm.
SEA DISTURBANCE ~~~ Rough. ≈≈ High.
BAROMETRIC CHANGE from 4h. to 7h. in tenths of millibars is entered beneath the temperature.

GENERAL INFERENCE.

A deepening depression centred off the Mouth of the Bristol Channel is likely to move southeast and then east. Cold wintry weather will continue with snow in many districts especially in the Southern half of Great Britain.

FURTHER OUTLOOK.

Wintry weather continuing.

Times of Issue of Gale Warnings.
▼South cone 0235 24th in Dists. 5 & 6
0835 24th in Dist 1.
▲North Cone 0840 24th In Dists 7 & 19.

Forecasts issued at 1025 G.M.T. G. C. SIMPSON, C.B., D.Sc., F.R.S., Director.

Fig. 90 *b*. Weather-map for the British Isles, with forecasts and general inference, on the day of the blizzard of 24 February 1933. The headlines from the London *Observer* on the 26th are inset. The map may be regarded as a note of introduction to the list of persons represented at the end of the chapter.

For readers of the Daily Weather Report the change seems to have been adumbrated or foreshadowed by the change in the maps of British weather on the larger scale for 7 a.m. on the 23rd and 24th February (figs. 89 *b* and 90 *b*) which are reproduced in order that the reader examining the situation may notice that in mapping the south-western extension of the area of low pressure the draughtsman has not used his discretion to round off the depression as Abercromby and other Victorian meteorologists might have done but, without any large array of coercive observations, has depicted an acute conflict between the winds represented by the isobars coming from the north and those which join up with the southerly winds of the east. No such conflict is indicated on the chart for 23rd February and only very mildly on the circumpolar chart for the 24th.

We have given, in general terms, the characters that take part in the drama as represented in the weather-map. There are of course endless varieties in the extent of their influence at any time, their intensity and their behaviour; but, speaking generally, from the behaviour since yesterday, with considerable experience a forecaster is able to draw a general inference as to the behaviour for to-morrow.

The primary inference is generally with regard to the distribution of pressure, and we may understand from Buys Ballot's law that the winds will be found approximately adjusted to the distribution of pressure. The forecaster associates with that distribution and its winds, anticipations of weather for the various parts of the whole area under his care. It would indeed be vain to attempt to give details of the practice beyond what is suggested by Abercromby's diagrams. We may, however, mention that cloud and rain are the most important elements from the practical point of view, and both of these depend not merely upon the plan of the atmosphere at sea-level which is expressed by the map but also on the vertical structure. To bring that into account requires a previous general idea of the distribution of temperature and water-vapour in the upper air and some knowledge of the actual conditions in the upper air at the time. Such information can be derived from observations of the height and character of clouds, from the evidence of wind-structure disclosed by pilot balloons and from observations of pressure and temperature at different heights by means of aeroplanes or balloons.

At present the information is fragmentary, much more inchoate than that which is available for the plan at sea-level. Some day it will perhaps be possible to enrich the Daily Report with vertical sections of the atmosphere and so to develop the idea of a working-model instead of the limitation to a horizontal section.

A good deal of organisation will be necessary before that is achieved and in the mean time the forecaster has to form his own ideas of the structure, very much in the same way as the listener has to form his idea of the map from the information about the barometer and what he can learn from his own instrument.

III. THE NORWEGIAN DUET FOR POLAR AND TROPICAL AIR WITH CYCLONIC ACCOMPANIMENT

Fronts and frontiers, a new departure since the War

In the general inference quoted on p. 242 we may notice the appearance on the stage of "small disturbances", new personages, so new indeed that their names are not in the official glossary. The language about them might have been used for secondary depressions and the change of practice is additional evidence of change of attitude towards the barometer-solo.

That development may be said to have begun with the publication of a work on dynamical meteorology and hydrography by Prof. V. Bjerknes and his colleagues which explained in English the proper way of representing movement on the atmospheric stage. That has been followed by the work on *Hydrodynamique Physique* by V. Bjerknes and others which is referred to in the Preface to this edition. In the interval V. Bjerknes worked upon the idea that the peculiar cyclonic isobars might really be the result of waves originating in the surface between two air-currents of different density and different travel, one warm, the other cold. The dynamics would be similar to that of the waves of the sea, and if the waves were set up in an interface that was not horizontal like the sea, the sea-level map would show a wavy line separating the dense cold air from the warm light air and the discontinuity already recognised at the surface would be explained.

The revision of the ideas of the weather-map introduced in Norway, is a new and notable departure since the War and indeed a consequence of the War, because it was engendered by the necessity for a careful

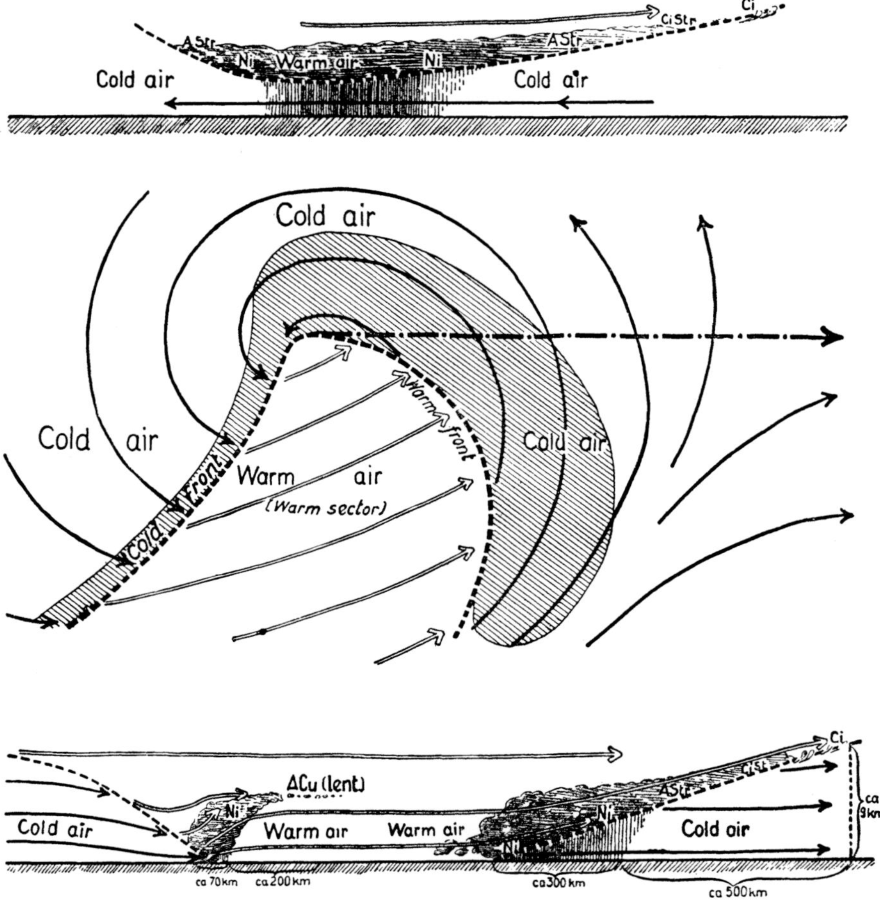

Fig. 91. Sketch of the structure of a cyclonic depression by J. Bjerknes.

In the middle a ground plan; the lines of **fronts** join at the centre from which the **steering-line** is drawn. Cross ruling shows the rain-area.

Above, is a vertical section, north of the centre; and below a corresponding section south of it. **Clouds** are indicated by shading and **rain** by vertical ruling. The motion of the air is shown by continuous arrows.

scrutiny of the meteorological information available in Norway when the telegraphic material for the ordinary weather-map was cut off.

We may introduce it with a diagram, fig. 91, of an idealised cyclone which must be the most widely known illustration of modern meteorology. It interprets the situation which is vaguely suggested by the discontinuities of fig. 88.

The first step was to recognise that when a depression was advancing eastward, the direction in which depressions generally do advance, a line drawn from its centre in the direction of the depression's progress and therefore called the steering-line was a line of discontinuity or of separation of two quite different kinds of air. On the north side there was air moving from the east characteristically cold and dry, and on the south side was air moving from south or south-west characteristically warm and moist. From that it was easy to understand why it rained in the front of a depression; if the southerly wind continued its course it would have to slope upwards over the top of the cold east wind, a condition in which rain is a physical necessity.

And then there are the well-known phenomena of the trough of the depression—a sudden change of wind from south-west to north-west with a change from perhaps rainy to clearing showery weather. This would be represented on the map by a wind from the north curling eastward round the centre and charging the inoffensive south-west wind in the military sense, driving it up overhead and again causing rain as represented in the plan of fig. 91 by cross-lines, and in the sections above and below the plan by vertical ruling.

Calling the southerly (south-easterly to south-westerly) wind equatorial or tropical air because it was warm and moist, and similarly calling the northerly (easterly, north-easterly to north-westerly) polar air because it was cold and contained little water-vapour, we obtain Abercromby's characteristics of the area of a cyclonic depression as appropriate to a region of conflict between tropical air and polar air. The line of front coming from the south along the trough passes through the centre, from there along the steering-line towards the future. The aggressive north-wester turning east may actually pass under the whole domain of the mild tropical wind and cut it off altogether from access to the centre on the ground. That is rather serious for the depression; its existence depends on the surface conflict of polar air with its opposite; it loses a necessary supply, the centre is technically "occluded" and the depression will perish.

Moreover, the division of the surface-air of the quarter of the earth's surface between lat. 30° and the pole into polar and tropical or equatorial air is much more general than is shown by a single depression. There is a regular supply of tropical air which can often be recognised flowing northward past Bermuda in the Atlantic, and turning eastward to bring warm moist air to supply the needs of the climate of north-western Europe; and if there be an intermittent supply of air from the polar region (as there appears to be), where the one supply meets the other the conflict will take place. History is not very communicative about what happens exactly to the victorious polar air but it is quite certain that it is responsible for the formation of a depression-centre and the depression-centres are strung on the "polar front", as the line separating the two airs is called, in most cases quite notably so. The battle of the Atlantic is not quite simple but undoubtedly wherever there is a depression-centre polar air or something like it is in conflict with tropical air or its equivalent.

It is the rotation of the earth which shapes these conflicts and the influence of convection that intensifies them and develops the rainy character of their weather.

So in making out the weather of to-morrow from that of yesterday and to-day, the modern meteorologist pictures to himself this conflict of tropical and polar air and bases the expression of his ideas upon a structure which is suggested but not actually expressed in the map at sea-level. It can, however, be gathered from information supplied by observations in aeroplanes and the study of the clouds.

The Norwegian mode

Instead of regarding a cyclonic depression or other group of isobars as a collection of states of weather which are somehow transported across the map and so affect successively the different parts of the region traversed, we have now to think of it as a region of conflict between polar air of cold origin and equatorial or tropical air originally warm with a line of separation between the two which marks what is called a front, a warm front if it is the tropical air which is advancing, a cold front if the advance is made by the polar air.

Taken round the northern hemisphere there are generally a number of examples of polar air advancing southward, and a line can be drawn between the recognisable polar air and the equatorial air against which in a number of cases it is advancing, or by which in other places it is being overridden. A connecting line will serve as an indication of the frontier or polar front in those regions where the differentiation is more difficult.

When a cyclonic depression is well marked by closed isobars the polar front or frontier line generally separates the area of the isobars into two very distinct parts, one in the occupation of tropical air as a warm sector separated from the larger remainder of the area which is occupied by polar air.

The plan of fig. 91 shows a cold front advancing from the north-west with a belt of rain and a warm front advancing from the south-west over the cold air through which rain falls, with the two fronts joining at the centre of the depression. The figure shows no isobars but if we surround the central junction of the two fronts by the line of an isobar which is crossed by the spiral current-lines going inwards, the area of the warm sector is clearly marked and is so labelled in the figure. With this arrangement it is the motion of the various fronts which the forecaster watches in order to define his anticipation of coming weather. The isobars may indeed be drawn to express what is in the forecaster's mind about the fronts but they are his servants not his masters.

And so the plan of the most recent meteorology is to form an opinion as to the origin, polar or equatorial, of the air-masses into which the map can be divided and to estimate the coming weather from their probable behaviour at the lines of separation. It is essentially a study of the character and behaviour of the surface-air, and as the surface-air is affected by contact with earth or sea and directly or indirectly by the sun's rays it is possible for an air-mass to modify its character if not to change its class. So we get maritime polar air, continental cold air, returning polar air and a whole scheme of phraseology to describe the

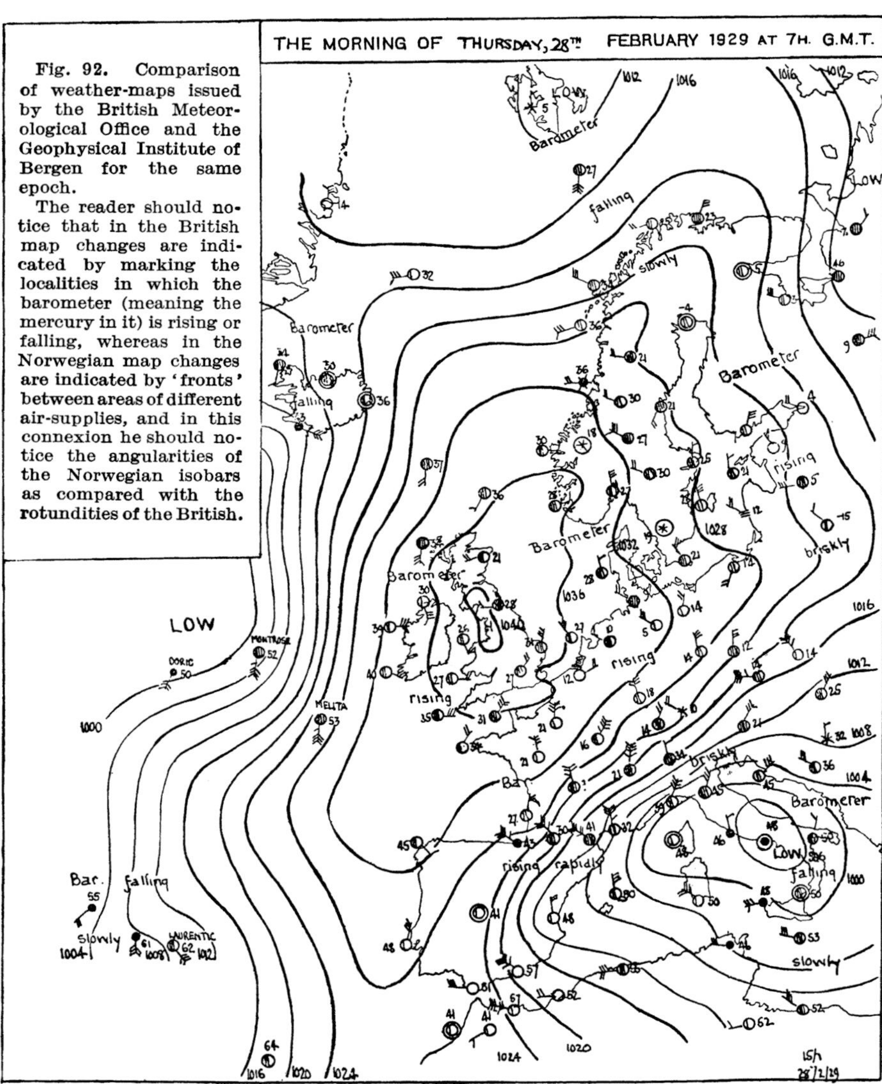

Fig. 92. Comparison of weather-maps issued by the British Meteorological Office and the Geophysical Institute of Bergen for the same epoch.

The reader should notice that in the British map changes are indicated by marking the localities in which the barometer (meaning the mercury in it) is rising or falling, whereas in the Norwegian map changes are indicated by 'fronts' between areas of different air-supplies, and in this connexion he should notice the angularities of the Norwegian isobars as compared with the rotundities of the British.

Fig. 92 *a*. Weather-map of the British Meteorological Office for 7 a.m. G.M.T., a day in the period of exceptionally cold weather in February 1929.

Showing the distribution of pressure by isobars, temperature in ° F by figures, wind by arrows, cloudiness by cross-lines in a circle and rain by black dots.

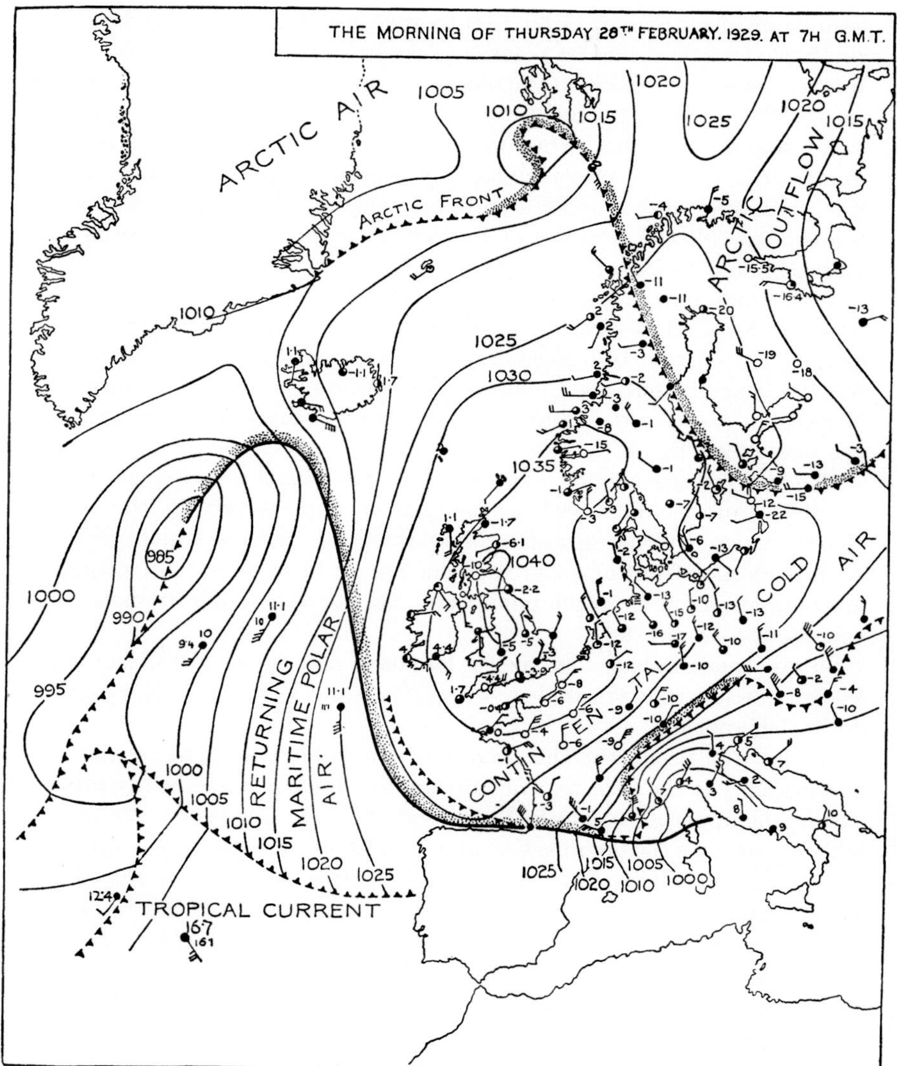

Fig. 92 *b*. Weather-map of the Meteorological Institute at Bergen for the same epoch as fig. 92 *a*, showing fronts by special device (see p. 253), pressure by isobars, temperature in degrees C by figures, cloudiness by a code of dots and rain by stipple.

behaviour of air-masses in relation to weather. The barometer loses its position as a soloist and its place is taken by tropical and polar air, and by their duet the weather of the temperate latitudes of Europe is brought into relation with the general rhythm of the world's weather.

Leading on from that, when weather for one part of the world can be expressed as the duet of two kinds of air meeting in conflict at the surface, it is natural for author and reader to ask whether they can find a similar solution for all the weather of humanity's experience however far it may be removed from the terrestrial poles.

We cannot pursue the subject far with the material available for a book of this kind because the most important features of the situation are generally details that need careful expression; but we may illustrate the difference of outlook by the reproduction of a British map and a Norwegian map for the same day—another February day, the 28th of the year 1929 (fig. 92).

The new practice finds expression in the weather-maps of many countries and is displayed perhaps most impressively in the daily charts issued by the Russian Meteorological Service which require both back and front of an area of six square feet, more than a square metre in all, for their accommodation. The corresponding British maps occupy two square feet.

So now meteorology uses a new language and as we introduced the Victorian solo with a vocabulary we may take our leave of the Norwegian duet with another.

We have noticed that the tracing of the behaviour is based on the observations at the surface. The information derived from the investigation of the upper air has to be incorporated by the forecaster. We await with expectation the organisation of a plan for representing the structure by a vertical section along one or more of the meridians of the map. Then the presentation of the drama will be no longer superficial; we shall be able to place the inference of the weather of to-morrow from that of yesterday and to-day upon a basis that will bear comparison with the achievements of the astronomers.

SOME PERSONS OF THE DRAMA IN THE NORWEGIAN DUET

Introduced in the International Section of the *Daily Weather Report* 1 March 1933 for the specification of air-masses which are taking part in the weather of the day. The symbols for fronts were revised as from 1 January 1937 (*see* p. 261).

Fronts or boundaries at the ground or sea-surface between masses of air of different origin are indicated, wherever their characteristics are well pronounced, by lines of a special kind. The motion of the fronts may be indicated by the winds at the surface but if necessary the motion of the upper air is taken as a guide.

A warm front is indicated by a pecked line. The air-mass which is moving towards this boundary is normally of tropical or sub-tropical origin while that which is moving away from it is normally of polar or sub-polar or maritime polar origin.

A cold front is indicated by a line of detached circles. The air-mass which is moving towards this boundary is normally of polar origin while that which moves away from it is of tropical origin.

Occluded front is indicated by alternation of dashes and circles. The region of air between the warm and the cold fronts within the region of a depression is of tropical or subtropical origin and is known as the warm sector. The air in front of the warm front and behind the cold front is of polar or maritime polar origin. During the life-history of the depression the air of the warm sector is lifted from the ground by the advancing cold front until the cold front has overtaken the warm front. The depression is then said to be occluded; the lines of the warm and cold fronts coalesce becoming a single front known as an occlusion front. It may carry on with a difference between its two sides although both are originally of polar origin.

Secondary cold front. A cold front in polar air following the first cold front. If the polar current is unstable there is sometimes an irregular series of secondary cold fronts of limited length accompanied by heavy squalls and sometimes by thunder. The fall of temperature is often due mainly to the precipitation and after the squall passes the temperature sometimes recovers almost its previous value. (From *The Weather Map*.)

The different modifications of air which take part in fig. 92 *b* are:

Of polar origin: **Polar air, Arctic air, Arctic outflow, Continental cold air,** Maritime polar air, **Returning maritime polar air,** Occlusion front.

Of equatorial or tropical origin: Tropical air, **Tropical current.**

The meaning is sufficiently indicated by the names and their positions on the map.

Fronts are indicated by toothed lines and the advance of a front by the direction of its teeth.

Occlusion or the locus of an occluded front is indicated by a thick continuous line. The thin continuous lines represent isobars.

On the British map isobars are drawn for steps of 4 mb, 1012, 1016, 1020 and so on; on the Norwegian map for steps of 5 mb, 1005, 1010, 1015. The reader should notice particularly the Norwegian isobar for 1010 mb in the far north, its behaviour on the coast of Greenland and in the "low" near Spitsbergen. The stippled areas indicate rainfall.

EPILOGUE

THE WEATHER-MAP OF THE FUTURE

Spoken from behind the scenes

From Nature's chain whatever link you strike
Tenth or ten thousandth breaks the chain alike.

A. POPE

We have seen that the theme of the Victorian meteorologists was the distribution of pressure. Wind and weather for any locality were indicated by its position with regard to the isobars. Little attention was paid to temperature or humidity as indicators of coming weather and none at all to actual discontinuity although in the phenomena of the trough-line and the V-shaped depression they were getting warmed to the idea.

It was in the Victorian age that the practice of extending the map over the land by reducing the pressure to sea-level was introduced. Numerically at any rate reduction increases the importance of pressure. It makes nice maps but not nice meteorology. Perhaps indeed it makes it easier to trace the long journey of a barometric minimum. The common opinion was that the distribution of pressure caused the winds, but no adequate provision was made in the scheme for the import or export of air which is represented for any locality by change of pressure.

We have seen too that in the Norwegian revision of the ideas of the play, the fronts which were brought face to face at discontinuities new or old became the leading actors. Pressure's isobars had to accommodate themselves to the suggestions of polar and tropical air on their march of adventure.

So now we find no longer pressure causing the winds but fronts with winds behind them, marking in a special manner the distribution of pressure if not actually producing it. Pressure remains reduced to sea-level over the land, still exaggerated in its measure but not in its importance.

And here we find a clue to some of the difficulties which the Victorians had to meet in tracing the travel of centres over land—we learn that most of them get themselves "occluded" when they come ashore. Their recognisable life is over, their part is played. What part the relief of the surface has taken in the occlusion is a question that some day the curious may fairly ask.

Then we have to ask for information about the regions behind fronts—what are they? and whence are they? They require as we have seen the use of qualifying adjectives and their energy is prodigious; but where they started from and who gave them their energy we are not told.

It has been pointed out that the surface-layer of air forms a very difficult field for scientific investigation, where the deceptive practices of nature are most abundant. And, further, it has been suggested that for the free air a thousand feet or more above the surface it is hardly possible to avoid the conclusion that the rotation of the earth provides that at any level the distribution of pressure shall adjust itself to the speed of the air-current with suitable high pressure on its right and low pressure on its left.

So for the weather-map of the future we need something that does not require pressure to be reduced, or in fact exaggerated, to sea-level, that uses actual temperatures and thinks of actual winds free from the impediments of the surface, where they came from, what made them and what is guiding them.

The weather-map of the future should have vertical cross-sections as supplements to the ground plan; the suggestion is supported by the classical representation of the structure of a cyclonic depression in Fig. 91, which gives a section south of the centre to show, on the left, cold air undercutting the warm air at the trough and, on the right, warm air making its way upwards over cold air to a height of "ca. 9 km", both operations producing rain as indicated in the plan. Also another section north of the centre to show warm air moving uneasily over a continuous cushion of cold air with an appropriate development of the rain area. There is no section through the centre itself.

For the Victorians the centre was the chief point of interest. The trenchant pictorial analysis of the Norwegian mode avoids it. And yet, it is a point of crucial importance because there the lines of the fronts join; warm air and cold air are on the boards of the stage together with the centre of the depression as a common point.

And let us not forget that great extension, upwards, of the surface picture of pressure is necessary in order to express the observed differences of pressure as those of the total weight of air per square foot or square centimetre; in reality their salient features may only be fully expressed in the upper air.

We understand that where warm air is brought into contact with cold air at the surface or in the upper air there will be a dynamical struggle by the cold air for the lowest place. And equally we understand that if in the struggle the warm air gets displaced upwards it suffers the inevitable cooling due to the real reduction of pressure that ensues. Saturation may be reached, rain formed, and the energy stored in the water-vapour becomes available for the exaggeration of the dynamical struggle until it may reach the dimensions of a thunderstorm.

And then we need not be surprised if the upper layers become the scene of the struggle and exhibit a tornado, a hurricane, or a depression with circular isobars as the result. A model of the structure thus evolved is a necessary addition to our mental furniture.

In Germany they are already producing maps of the sky compiled from local observations of cloud with which Abercromby's prognostics (fig. 85) might be compared.

On the real stage of the weather-map there are quite a number of distinguished actors; first the sun who manages the whole display, next the earth which sets the stage, disposes the energy and by its rotation leads the dance, next gravity, the wind-raiser, who conducts the performance, with a special reserve of energy in the water-vapour, and finally the atmosphere in which temperature and vapour, pressure and wind join in the drama with a symphony which we have to express in the weather-map of the future without any break in the chain with which Nature links them all together.

Years ago when the weather-map was young the leading idea was that in course of time so much would be learned about the air and its ways that we should catch the movements of the symphony within the twenty-four hours between the daily maps and then for longer and still longer movements with knowledge slowly broadening down from precedent to precedent. But the development has gone the other way. An evening map was called for by the newspaper press, and then a mid-day map for harvest forecasts. And now four maps within the day are published and others are interspersed for the service of the air.

Yet for meteorology as a science the original idea still has its charms, and when we can really identify the structure of the symphony they will be recognised.

The play is there for all the world to see and hear; and the action will be plain when our minds can get the true perspective into focus.

CHAPTER VI

WHERE OUR RAIN COMES FROM

We run because we must
Through the great wide air.

SORLEY, *The Song of the Ungirt Runners*

BRITISH RAINFALL AND SOME OF ITS FEATURES

For presenting the play of pressure, wind and weather in Chapter v we have made use of the Weather-map to display its facilities for forecasting in the forms which we have called the Barometer Solo and the Norwegian Duet, because in the modern world those aspects of the drama have claimed more attention than any other; but turning over the pages of our own version other aspects present themselves which may perhaps deserve attention in the weather-maps of the future.

We may notice for example that the recognised elements of the map, the depression with its secondary and the anticyclone, or their extensions, the trough of low pressure or the ridge of high pressure, are said to be moving and one naturally wants to ask what it is that moves, when the air, with the water which it carries, is the only constituent of the picture. Should one regard the weather-cock or perhaps the clouds as the criterion of the motion?

For the barometer solo of the nineteenth century a group of closed isobars of pressure, cyclonic or anticyclonic, was recognised as a traveller. Indeed until well on in the present century closed isobars were the *dramatis personae*, with hurricanes, typhoons and tornadoes as demonstrative examples. For the Norwegian duet the motion upon which attention is concentrated is the advance of "air-masses" across lines of discontinuity where separate air-masses are in contact.

Years ago physical geography as taught in schools had no difficulty in assuming that the effect of the solar radiation upon the earth's surface and atmosphere was to cause the differences of pressure which in their turn caused the winds. I remember a schoolmaster whose special subject

Fig. 93. NORMAL MONTHLY FALL OF RAIN IN THE BRITISH ISLES

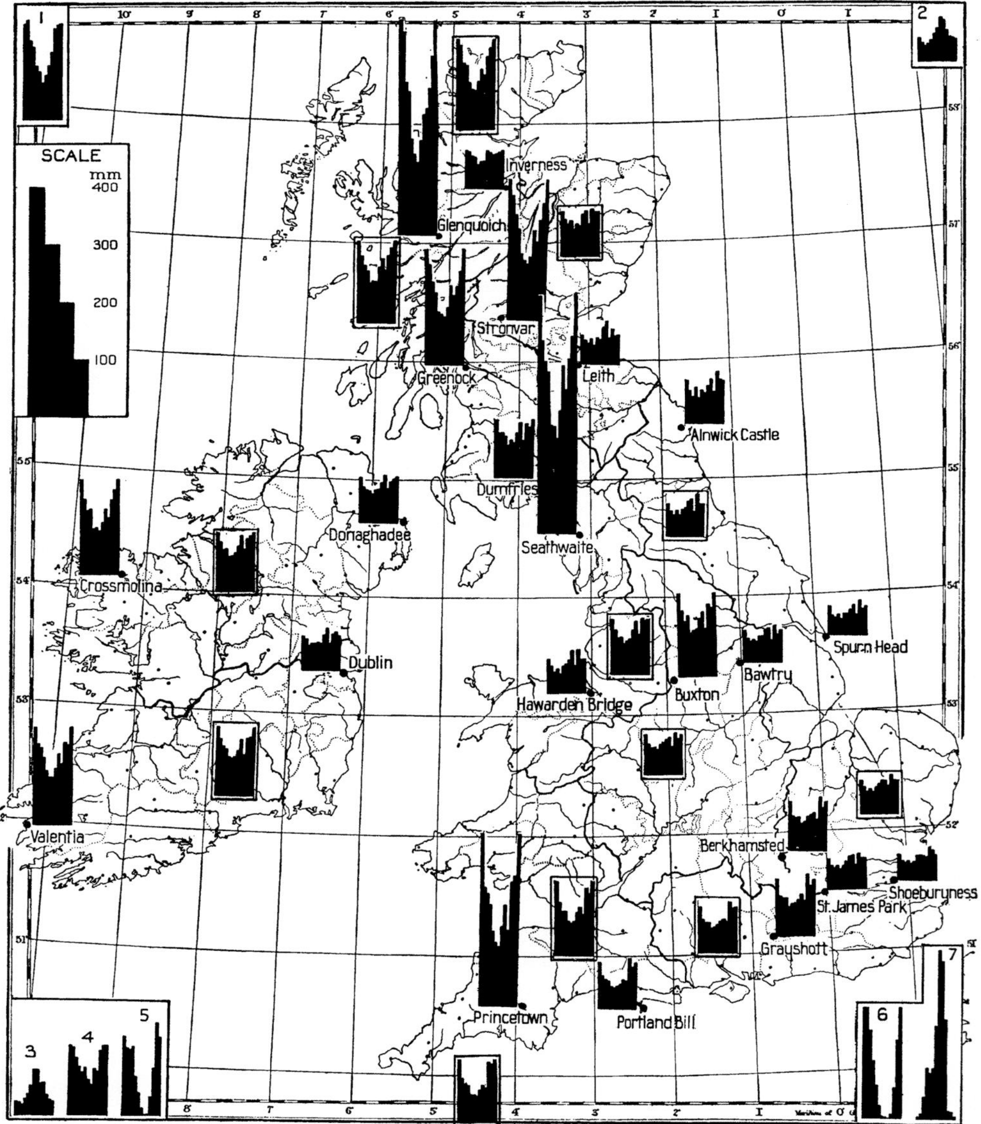

was geography being visibly shocked when, as a pupil of bygone years on a visit, I demurred to the statement that wind was the flow of air from high pressure to low. Indeed to-day, recognising the various processes by which currents of air may be produced, we may be pardoned for asking whether it is not the natural winds which cause the various features of the distribution of pressure.

Looking at the matter from a general point of view we may recognise wind and rain with associated temperature and the state of the sky as the chief features of weather referred to at the head of the Watchers' chapter on p. 79; and in this connexion we may remember that the moving air of wind always carries with it an amount of water which depends upon its experience as regards temperature during its travel, and that leads us to think of the transport of water as one of the features of our meteorological system.

It has been remarked for example that British rainfall is a West Indian product, and we may use that remark as an invitation to look into the question of the source of the water which falls upon the British Isles as rain.

As an introductory presentation of the problem we may use a map (fig. 93) of the monthly distribution of rainfall over the British Isles which appeared in a recent edition of Vol. II of the *Manual of Meteorology*. The information which it includes is based on the material compiled years ago for the *Weekly Weather Report*. It presents the normal monthly fall for each of the twelve districts into which the country was divided: Scotland N., E. and W.; England NE., E. and SE., Midlands, England NW. with N. Wales, and SW. with S. Wales; Ireland N. and S., and the Channel Isles with Scilly; and there are also for each district corresponding figures for the normal rainfall of two stations, chosen respectively as having the highest and the lowest rainfall of the district.

Legend for Fig. 93.

Normals for the twelve districts are given by diagrams enclosed in frames. Normals for the stations of highest and lowest rainfall in each district are given by diagrams, the points indicating the positions of the stations. Data from the *Book of Normals*. Each diagram begins and ends with the normal fall for the month of December. In the four corners are examples of corresponding normals for: 1. Thorshavn of the Faroe Islands in the North Atlantic, 2. Moscow in the great Eurasian continent near the Asiatic border, 3. Winnipeg in central Canada, 4. Horta in the Azores, 5. Gibraltar at the western end of the Mediterranean, 6. Beirut at the eastern end, and 7. Addis Ababa under the equator at the source of some of the water that feeds the Nile.

These examples at once suggest that places may be distinguished by the peculiarities of the normal distribution of their rainfall between the months of the year. Rainfall is especially a summer phenomenon at Winnipeg, Moscow and Addis Ababa, and a winter phenomenon on the Atlantic islands and at Gibraltar and Beirut as well as at the stations of the western seaboard of Great Britain and Ireland; and in our map of the rainfall of the world (fig. 62) we see at once that summer rainfall is characteristic of the interior of the great continents whereas at the stations which are within easy reach of the oceans winter rainfall is characteristic. That is hardly what one would expect and so raises the question of where the water comes from.

It might possibly be suggested that the summer growths of trees and herbage are able to supply water for their own neighbourhood by using the circulation of the sap to extract the water from the store in the soil and put it into the air through the summer foliage. Certainly a vast quantity of water gets into the air in that way: "An acre of cabbage will use two million quarts of water in a season and 200 beech trees on an acre require nearly double that amount. One of these trees loses about 80 quarts of water as vapour daily from its leaves." (D. T. MacDougal quoted in *Bulletin of the American Meteorological Society*, April 1932, p. 80.) That would hardly explain the rainfall conditions of Palestine where drought in summer is conspicuous. For example, at Jerusalem the monthly rainfall for 1935 is reported roundly in millimetres as: January 102, February 221, March 23, April "drops", May "drops", June, July and August 0, September "drops", October 36, November 31, December 17. Year 430. We need not be surprised that a thunderstorm in Harvest was once regarded as an expression of displeasure on the part of the Almighty and that the forecaster was regarded as a prophet.

Our British rainfall cannot be treated exclusively as either a summer or a winter phenomenon. We are not in the middle of a continent, we have coasts which face all points of the compass with considerable variation of level of the land enclosed. So we ought not to fail in thinking of the association between the rainfall which is shown on our charts and the winds by which it is brought to our locality.

Our weather-charts tell us something about this. Let us take for example (fig. 94) those for Midsummer Day and Christmas Eve of 1937 as representing conditions for summer and winter. We may find in the charts of the International Section of the Daily Weather Report the distribution of pressure, temperature, wind and weather at the hour of the chart, and the fronts of the air-masses which are moving over these islands with the nature of the air that they bring, in accordance with the

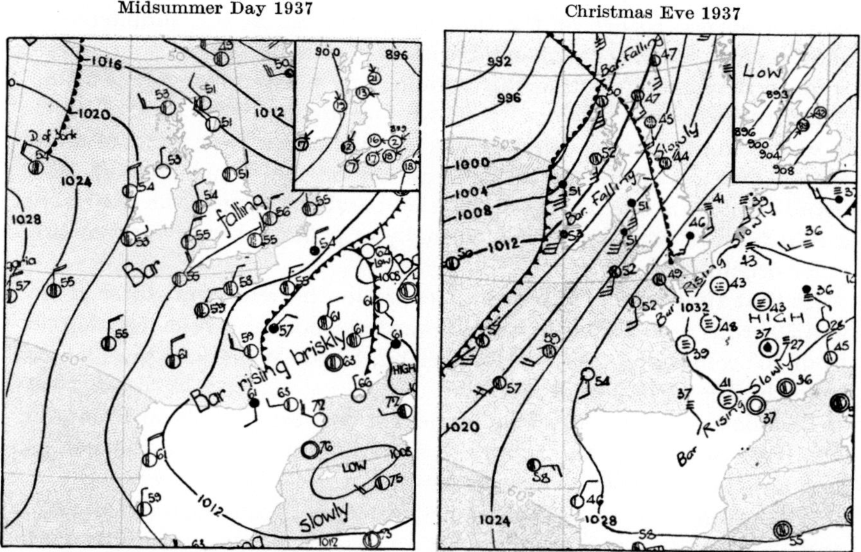

Fig. 94. Wind, temperature and weather from charts for 7h. 24 June and 24 December 1937 with fronts. Also small inset charts to show the motion of air at the height of 1 geodynamic kilometre (3340–3350 feet). A black dot indicates rain falling.

Norwegian suggestion set out on p. 253 except that the method of representing fronts on the British map has been changed. A warm front is now marked by a succession of half-circles and a cold front by a succession of black teeth, while an occluded front is shown as an alternation of the two.

For the summer map we find some cloud indicated but no rain for these islands. Air is flowing from the North over England and Ireland with a westerly component in the far north. A warm front is aligned north and south, tropical air of the Norwegian duet is advancing eastward towards Scotland and an occluded front facing both ways is displayed within the range of France and the Netherlands.

And these charts have for us the great advantage of giving an indication of the temperature of the water of the ocean by means of greater depth of the blue colour for water above 60° F. with its boundary line landing at the corner of the Spanish peninsula in the summer map, showing a region of temperature between 60° and 50° up to a line running from west to east near the 60th parallel, and below 50° north of that.

For the winter example of fig. 94 we find strong south winds forming a warm front, that is tropical air, advancing eastward across Great Britain, with a diversion to the west off Scotland and Ireland, itself invaded, along a line just west of Ireland and reaching to the Azores, by a cold front of which the air is normally of polar, sub-polar or maritime polar origin.

At that time of year the line of separation between the water of 60° and 50° runs across the sea ESE. towards Lisbon, and that dividing 50° from 40° comes down SE. across Ireland, South Wales and the Channel from the Isle of Wight to the Seine.

To follow out the idea of the connexion between the Warm Front and the West Indies we must appeal for information about the general circulation of air over the earth's surface.

The introduction of a cold front with a line of advance sideways-on against an air-mass moving from the south suggests the possibility of the cold air advancing by pushing itself under the warm air. It might in that way help to rob the tropical air of some of its water. The rainfall indicated at two points on the west coast of Ireland serves as a reminder of that kind of possibility though the dots of like significance on the coast of Wales seem to provide rainfall in the middle of the warm air-mass. The possibility of difference of origin of the air at different levels should not however be disregarded. The Upper Air Section of the Daily Weather Report tells us about that and we have reproduced what is represented on the maps of the published reports about the conditions at 1000 geodynamic metres (3340–3350 feet) in insets to the two charts; but to deal with the matter fairly we must formulate our ideas of the way in which the air moves in currents over the surface of the globe.

THE ATMOSPHERIC CIRCULATION

Accepting the style and titles suggested for the identification of the different kinds of air that make our weather—tropical air, polar air, arctic air, maritime tropical air etc.—we may form some idea of the circulation of air over the globe, which brings air from the various sources for service at the fronts as required; and for this we must rely primarily on observations of wind over the oceans which have been accumulated in "Marine Divisions" since the beginning of international meteorology in 1854, which happens to cover the birthday of the present writer. They provide material for wind-roses on monthly charts of the various oceans and thus amplify the information given in Halley's famous map of the winds. The corresponding observations on land would be at different levels. Meteorologists have the audacity to reduce readings of pressure and temperature at land-stations to sea-level in order to extend the isobaric and isothermal lines over the land; but our audacity does not extend to reducing the observations of wind to sea-level, because different features of the circulation may be in control at different levels.

According to Buys Ballot's law as set out on p. 235 the lines of flow of the air over the oceans may indeed be expected to correspond moderately well with the run of the isobaric lines; so we get a good idea of the general flow of the surface air as moving along the isobars represented in the maps of fig. 51, always keeping the area of lower pressure on the left in the northern hemisphere and a little biased towards a drift from the high to the low at the surface in consequence of the local friction between air and water.

The combination of observations for different parts of the oceans is summarised for January–February and July–August in the hemispherical maps set out in fig. 53, which are indeed based on maps on Mercator's projection set out by Professor Köppen in Bartholomew's Atlas.

The main idea which we may carry away from a study of the maps is one that was familiar in the days of physical geography, namely a flow of air from north or south towards the equator, deviating more and more to the westward as the equator is approached and constituting what were known to seamen as the North-east Trade-winds in the northern hemisphere and the South-east Trade-winds in the southern. The maps do not show them flowing, as we used to think, towards the equator uniformly from all longitudes but rather as forming a circulation round areas of high pressure about lat. 35° N., with a centre in the Atlantic and another in the Pacific more obviously developed in the summer than in the winter; but so far as the equatorial region is concerned the tendency seems to be to flow along the equatorial zone from east to west over the Atlantic and Pacific with an exceptional flow eastward over the Indian ocean in the summer, the last forming the south-west monsoon.

At first sight the most striking feature of Köppen's maps is perhaps the belt of short black arrows in the northern region of the great ocean that surrounds the Antarctic continent, a belt of strong winds, force over 6 on the Beaufort scale according to the legend, but not at all "steady" judging by the shortness of the representative arrows. That is the belt known to seamen as the Roaring Forties, which seems to be made up of boisterous eddies travelling eastward from the west. The belt is a little more compact in January–February, the southern summer, than in the winter, July–August.

The winds which are epitomised by the black lines might perhaps be reasonably represented by a wind-rose of a type not very different from that for Scilly (fig. 66) making due allowance for the normal direction of the most vigorous winds. And perhaps the transition from the forties to the fifties of southern latitudes might find a pattern for its wind-roses in that of fig. 68 for the region of Cape Horn.

In the map of the northern hemisphere we may also find unsteady winds which appear to be making an effort to establish a circumpolar circulation, but the distribution of the land spoils the effort and the shortness of the arrows obviously indicates an interference of winds from other directions with those marked by the arrows as coming "normally" from the west or south-west but very unsteady.

If, however, we are thinking of winds which are marked on the maps as steady instead of the conspicuously variable ones, we can appreciate the flow of air associated with the wind-roses of fig. 65 and representing the features of the South-east Trades throughout the year in the neighbourhood of St Helena and in the same belt of latitude near the Brazilian coast. And we may also cite the steady winds in opposite directions in summer and winter represented in fig. 67 as typical of the behaviour of the monsoon area of the Arabian sea.

Lines of steady winds, continued with only slight alteration of direction, imply long distance of travel, and we should like to regard the local origins of the steady winds as marking the commencement of circulations of air over the globe. It would be necessary to supplement the information which the map shows by some suggestion as to the cause of the original steady drift and the arrangement by which the travelling air is replaced as it leaves the region of its first appearance on the map.

We ought to be able to trace a connexion between the steady winds flowing towards the equator and the groups of arrows which show a flow from west to east over the oceans but by the brevity of their arrows they show that they have lost the steadiness of the westward flow. They are sufficiently well-marked to account for the carrying of a good deal of water from the tropical zone to the western coasts of the great continents, and this is the special point of interest for us at the moment, because the British Isles are in a region where the average flow is marked as one of unsteady south-westerly winds on the charts.

We may thus picture to ourselves a sub-tropical anticyclone between latitudes 30° and 40° with a steady flow of air from East to West on the southern side and an irregular flow eastward on the northern side of which the roaring forties are typical in the southern hemisphere. We can attribute the greater regularity of the South to the absence of interfering land-areas below 40° S.; and, recognising high land-areas as possible localities where the gravity of nature can express itself as initiating an air-current over the surface, we may attribute the want of steadiness in the westerly current north of the anticyclone as due to the interference of winds from the northern areas becoming involved with the westerly and developing centres of low pressure by the winds in the region beyond the overpowering influence of the westerly. It is, of course, a peculiarity of living on a spheroidal revolving globe that North and South are not such obvious lines of travel as one might suppose, and the motions of the air that conspire with the air moving East or West to produce or accompany depressions are North-Westers or North-Easters; and, in like manner, the motions which originate as South may come on the map as South-Westers or South-Easters.

The air which feeds the westerly current can generally be identified in the neighbourhood of Bermuda in Lat. 31° N. as being usually supplied from the surplus of the intertropical air that finds its way northward, and it is on that account that the supply of air carrying water to Great Britain can be associated with the West Indies; and the idea is reinforced when we recognise Bermuda as the turning point of the course of tropical hurricanes that begin their travel with motion westward but ultimately arrange to go east with their violence somewhat modified by the change in their air-supply as they travel.

The motion from the east below latitude 30° N. and from the west above latitude 40° N. with the interference of the north-wester or the north-easter is the kind of general circulation which we have in mind as the mode of supplying tropical air to the temperate zone. We may therefore pass on to consider the details of some parts of the irregular circulation as exhibited upon the daily charts of weather.

We need not regard the easterly or the westerly as the only persistent winds. We learn from experience in England that the air may blow from the continent east of us for days or even weeks, and in 1938 we have learned that a current of surface air from the north, north-east or north-west, may be persistent for a long time and be associated all the time with a current in the same direction in the upper air at 1000 or 2000 metres.

We may remember that in the autumn of 1937 the Russian scientists were carried on their ice-flow southwards from the North Pole to the coast of Greenland as indicated on our frontispiece, so that during their travel the north winds extended over a wide range of polar latitudes; and their experience suggests that the polar regions find accommodation for some part of the general circulation which has not yet been fully identified and which provides a regular supply of air to form the northerly component of cyclonic depressions and to feed the drift southward of the air-flow of lower latitudes towards the equator.

THE RELATION OF RAINFALL TO THE GENERAL CIRCULATION

In what precedes we have been thinking of the motion of air along the surface and considered it possible for the flow of air towards the equator to be compensated by flow from the intertropical region to the region of variable winds north of the tropical anticyclones of the oceans. That would not have been regarded as reasonable in days gone by. Sir Francis Galton defined an anticyclone as a region of descending air which supplied the superficial requirements of weather by flow from the base, and in physical geography we were taught to regard the surface air as flowing towards the equator, rising there to the upper levels on account of its warmth, and being carried by its own overflow to form the tropical anticyclones which were the reservoirs of surface air for meteorological purposes.

By way of distinguishing between these aspects of the atmospheric circulation let us consider some examples of the distribution of the meteorological elements and their relation to rainfall displayed in our Daily Weather Reports.

We may begin (fig. 95) with the part of the map that includes the intertropical region of the West Indies, though it is separated from Britain by the whole of the Atlantic Ocean. Our record dates from 19 October 1937 and shows a rainy evening (the map is for 8 p.m. 75th meridian time) in the eastern United States lying between a High over the western Atlantic and the Rocky Mountains.

If we can regard the exterior isobars of the High as a general indication of the motion of the air, and the suggestion is borne out reasonably well by the indications of the winds at various stations en route, we may recognise a flow of air below 30° N. latitude towards the Caribbean Sea and still more definitely a flow northward from the Caribbean Sea along the Mississippi valley to the Great Lakes of North America and even to Labrador and Hudson's Bay.

This represents a very interesting type of American weather that enforces the suggestion that the water which maintains the flow of the Mississippi to the Gulf of Mexico has to make its way from the Gulf to the sources of the river before finding its way back to the Gulf.

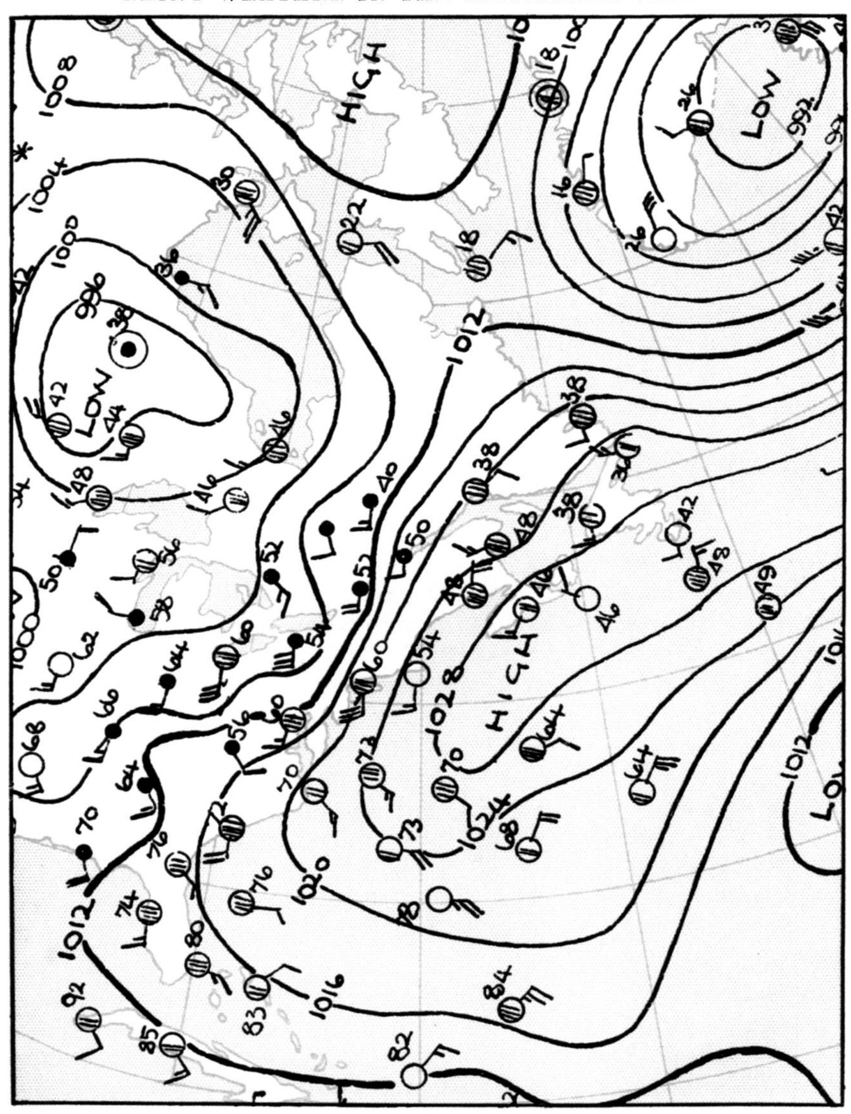

Fig. 95. From the Chart of Weather in the Northern Hemisphere, Daily Weather Report, 19 October 1937.

A similar relation between the air-flow upstream and the river which carries the water back to the sea may perhaps be recognised in the case of the Amazon but not in the case of the Nile.

The amount of water involved in the North American exchange is very considerable as the moist air moving upstream is attacked on the side by cold air from the Rocky Mountains, and the upper regions of the Mississippi are sometimes flooded thereby, as in the very notable floods in January and February 1937 of which an account was published in *Nature* of 16 October 1937, with a statement by C. L. Mitchell, forecaster, U.S. Weather Bureau, in an article on "The Ohio-Mississippi floods of 1937" by R. W. Davenport, U.S. Geological Survey.

The extremely heavy rainfall over the Ohio valley, Tennessee and Arkansas and part of the adjoining areas was in general caused by the fact that this area was so located with relation to the very deep areas of high pressure on either side that at the earth's surface the line of contact between the warm, moist air from the south, and the dense, cold air of polar origin that came in over the Ohio and middle Mississippi valleys on many days from the north and north-east, lay somewhere over this area much of the time; and the less dense warm air from the south (or south-west) was forced to rise over the cold and denser air. The rapid lifting of the very moist air of tropical origin resulted in abundant precipitation.

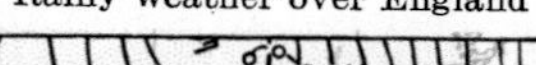
Rainy weather over England

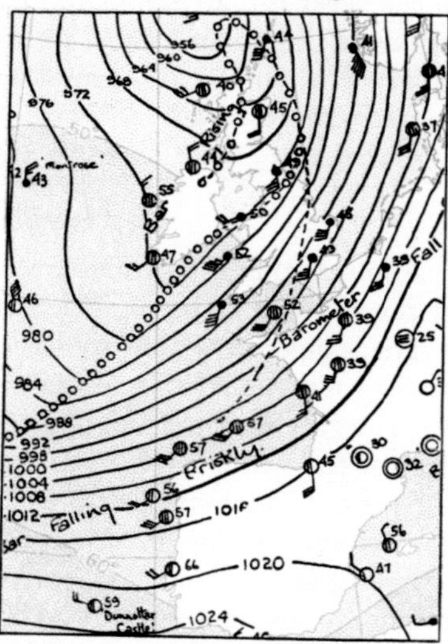

Fig. 96. Chart for 14 December 1936. A south-westerly current of tropical air with polar air advancing from the north-west.

We might cite also examples of maps which suggest that the rainfall in the eastern United States is supplied by air from the Gulf of Mexico; but when we consider the rainfall of the British Isles the connexion is not so obvious.

As an example of rainy conditions in England let us look at an extract from the map for 7 a.m. on 14 December 1936 (fig. 96), which exhibits a set of a dozen isobars curling round from south-west to north with a pecked line crossing them indicating, according to the earlier convention, a warm front, followed by a line of small circles indicating a cold front, and terminating the area of wind from the north impinging upon the warm south-wester. Rain is shown to be falling at seven stations in the British Isles—at the Shetlands and at Scilly, at Holyhead and Pembroke in the west and at Kew, Gorleston and Tynemouth in the east.

The analogy of these conditions with those of the map representing rainy weather in the Mississippi valley is really rather striking, if we can regard Greenland as providing a supply of cold air from the polar regions to invade the warm south-wester instead of that which came in from the north and north-east over the Ohio and middle Mississippi valleys, and if we can regard the south-wester as having travelled over the Atlantic we may reach the conclusion that Great Britain shares with the United States the air which is supplied by the intertropical flow of air from east to west and is diverted northward by the conditions of motion originating in the West Indian region and travels sometimes up the Mississippi valley as far as the Great Lakes, providing the water for the Mississippi, while at other times the air or another part of it travels northward along the western shore of the Atlantic and supplies the westerly winds on the northern side of the Atlantic anticyclone.

The effect of temperature.

Allowing that as a probable condition, we may now consider the question of the amount of water which the air can carry with it on its travels.

The temperature that we have to consider in relation to such a question as the source of British rainfall may perhaps be summarised as related to 80° F. as the temperature of intertropical waters, with 60° as the temperature of the water on the southern European coasts, 50° as the temperature of our northern waters and 40° where one begins to consider Arctic conditions. We have already indicated the positions of the variable temperatures in summer and winter in the temperate latitudes with which we are concerned and we can obtain the information which we require from the table of p. 104.

We may take 80° F. as a sort of normal temperature for the surface air of the tropical zone, and at that temperature when saturated the water-vapour would have a pressure of 35 millibars, a density of 25·4 grammes per cubic metre, and a kilogramme of dry air might carry approximately 22 grammes.

If in the course of its travels the air got cooled to say 60°, the pressure of the water-vapour would be reduced to 17·7 mb, about one-half, its density to 13·3 and the load carried by a kilogramme of air to 10·9, again about one-half. And going further with a reduction to 40° the pressure of vapour would be unavoidably reduced to 8·4, its density to 6·6 g/m^3 and the load on a kilogramme to 5·2. Eight degrees further would carry it to the freezing-point of the water and we have to think of particles of ice instead of drops of water; at 30° the charge of water-vapour would have a pressure of 5·6 mb, a density of 4·3 g/m^3 and a load of only 3·4 g upon a kilogramme of dry air.

We must think in passing of the ways in which the travelling air might lose its temperature on its journey to the temperate regions. It may be lost automatically by radiation if there is any region within sight where there is a mass at a lower temperature, and still more if it is exposed in a clear sky. It loses temperature also if it is lifted or otherwise moved by any process to a position of lower atmospheric pressure, or if it somehow gets mixed with another sample of air of lower temperature.

The fall of temperature may be arrested if the reduction of the moisture takes the form of condensation of some of the vapour into drops, water or ice; for then what we call the latent heat of the vapour is available for the environment of the air in which the condensation takes place, and then are brought into consideration all the physical properties of water in its three stages of vapour, liquid and solid.

We can however infer that tropical air starting from the equator and reaching the shores of the northern latitudes can only carry the water appropriate to the temperature of the air at that part of its journey.

The actual tracing of the changes involves the comprehension of all the stages in the progress of the air along its path over the globe.

Since air can be and is in practice most heavily loaded with moisture in the equatorial region, it is natural to regard that region as a possible origin of the rainfall of the temperate zones; and the association of the travelling air with currents from the west is suggestive of the West Indian seas as a probable source of supply of condensable water for the British

Isles by air which may travel directly as part of the westerly current, or pursue some more devious course and appear off our shores as coming from the south or south-east. It is easier to explain British rainfall that way than to give a definite explanation of rain falling from winds of polar or continental origin. Still the flow of air over the North Sea is sometimes reasonably held to account for drizzle over the Eastern counties, and the map of 3 January 1938 showed a number of stations in a north-south line along the country with rain and a north wind; and when the winds of the upper air were scrutinised to find out if the warm south-wester was being undercut by the north wind at the surface, it was found that at the 1000 and 2000 metre-level the wind was still north.

Local rain may be produced in wind from any direction if after loading itself to the limit with moisture it finds itself impelled by a cooler neighbour or some other obstacle to make room at the surface and betake itself to the upper regions where it cools automatically.

So we have still something to learn about the details of atmospheric circulation which ought to be easier of accomplishment in these days when so much knowledge of the upper air is at our disposal if it can be suitably expressed.

SOME TYPICAL WINDS AND THE ENERGY THAT THEY CARRY

We have cited as the most general feature of our weather-maps the flow of air from the east between latitude 30° and the equator, gradually drifting southward as it goes forward to feed the equatorial current, and a corresponding current from the west, above latitude 40° or 45°, of "tropical" air returning from the equatorial region. The air returning from the west helps to develop the phenomena of our weather on its northern border and with the assistance of currents from the north-west or north-east to form cyclonic depressions instead of steadily moving air.

This combination develops some striking features of our weather-maps of which we should like to display some examples. But we may preface these with an example of unusual persistence of the east and west currents, with an extensive development of the anticyclone between them, that gave our month of March 1938 the title of the warmest and sunniest month of that name for some 138 years. As the day chosen was 4 March, the map (fig. 97) may perhaps be regarded as a birthday gift to the author. It is peculiar in respect of the representation of a definite flow of air northward on the western side of the anticyclone over the middle latitudes of the Atlantic. The flow of air from the west is indicated

An obstinate anticyclone

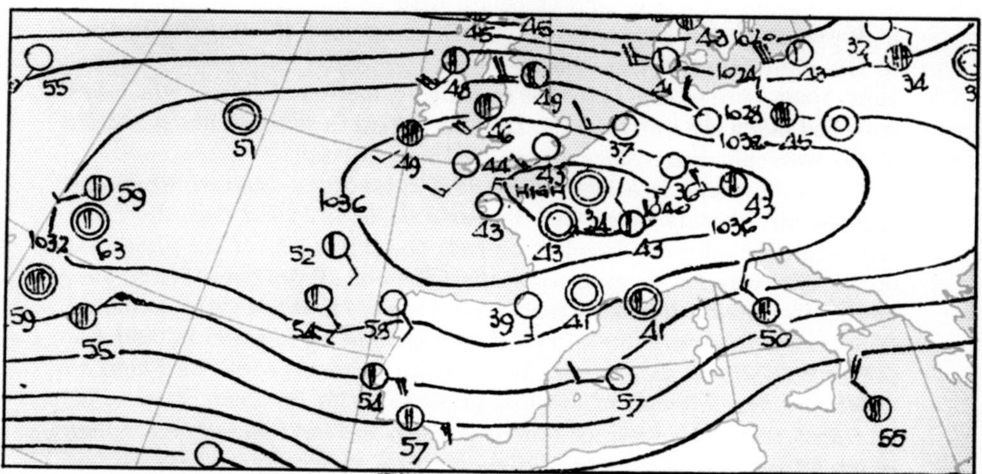

Fig. 97. Chart for 4 March 1938 showing an anticyclone that persisted with little change during the month, illustrating the British Rule of the Road, and also, if it be true that air moving on a spherical or spheroidal surface has to keep a little to the left if it wants to go straight, it may go to show that two currents from east and west passing one another in parallel lines under the influence of the earth's rotation must produce a high pressure between them. The cause of the anticyclone in the illustration may be the motion of the two currents. North winds and south winds are transitional.

by four isobars above lat. 45° while other four, below that latitude, indicate the supply of easterly winds.

We are citing this example of a current of air from the east in conjunction with a current north of it from the opposite direction because it affords an effective illustration of the suggestion, made some years ago and already referred to in this volume, that it may be the motion of the air which causes the distribution of pressure instead of the converse.

We know that owing to the rotation of the earth an object projected in any direction from a point on the earth's surface will be diverted towards the right of its projected path in the northern hemisphere. The air starting from the west will develop a drift southward while that from the east will develop a drift northward, each increasing as the track is pursued, and in order to keep a straight path over thousands of miles for some of the air the part nearest to the air moving in the opposite direction must be diverted from its course and associate with its opposite to form a local high pressure with gradients which will counteract the impulse to turn to the right. So it seems reasonable to regard the anticyclonic isobars 1032, 1036, 1040 as *caused* by the continuous motion of the air which has travelled along the isobars 1028, 1024 on the south side and along isobars of the corresponding values on the north.

This anticyclone is a case which illustrates the procedure when the original air-currents follow the British rule of the road and allow the current in the opposite direction to pass on their right. If the directions of the two currents are reversed so that the one allows the other to pass on the left the rotation of the earth would require that the currents should be diverted from the line of separation instead of towards it, and the pressure would be diminished. Whether an example could be found in which the development of the low pressure to its final condition could be traced on similar lines we cannot yet say. We have to remember that in fig. 97 increase of pressure causes increase of density and increase of temperature; but in the other case diminution of pressure causes reduction of temperature which may take the air below the condensation point, and all the physical consequences of the condensation have to be considered. The anticyclonic stability is replaced by cyclonic instability.

We have used this bit of the map to illustrate the formation of what we may call an obstinate anticyclone because there is a suggestion of persistence in all its features; and now we may give a little attention to the energy which is stored in the moving air. The classical expression for the energy is *vis-viva*, defined in the *New English Dictionary* as "the operative force of a moving or acting body, reckoned as equal to the mass of the body multiplied by the square of its velocity".

So far we have been considering the moving air as a water-carrier, and we have noticed on p. 236 that its main load is the energy of its motion which must take its part in the dynamical arrangements of the globe.

From the fact that the velocity necessary to maintain the pressure-distribution is estimated by the formula which expresses the pressure-gradient dp/dx as equal to $2\omega v\rho \sin \phi$, where ω is the angular velocity of the earth's rotation, v the velocity of the moving air, and ϕ the latitude, it follows that the kinetic energy of the wind over any bit of the map represented by a square lying between consecutive isobars with a pressure-difference b and with a height h of its own, is expressed as $\frac{1}{8}\frac{hb^2}{\rho\omega^2}\operatorname{cosec}^2 \phi$. It will be the same for any one of the squares between consecutive isobars no matter how wide apart or near together they may be. The reader will forgive the quoting of a formula for this vital relationship.

On the map of fig. 97 that we have been considering, the scale is approximately a forty-millionth of the natural scale, so that when the distance between isobars is 4 mm. on the map it represents 160 kilometres, or 100 miles, of the earth's surface, and the energy of an air-current 10 metres thick covering a square between consecutive 4 mb. isobars should be fourteen million kilowatt-hours, using the B.T.U. (Board of Trade Unit of energy for regulating the supply of electrical energy) as its numerical expression.

As a unit for comparison we may consider the energy of a motor car, a ton in weight, travelling at the limiting velocity of 30 miles an hour, approximately 15 metres a second. The number of such cars necessary to carry the same amount of energy as the 10-metre thickness of wind over the square is 600 million. On the map that we have been considering there are strips between consecutive isobars which may extend over 2500 miles and contain 32 squares; the energy of such a strip would be equivalent to about 19 thousand million of our motor cars.

We are familiar with the energy of moving bodies in many other forms and with these the energy of the wind over a single square between consecutive isobars may be compared. We have used the travelling motor car, supposed to represent a ton moving at thirty miles an hour, as a familiar example. Energy takes many forms and its measures may be expressed in various units; we give here a little table of examples with their expression in the fundamental unit of energy of the c.g.s. system, the erg.

EXAMPLES OF SOME UNITS OF ENERGY WITH THEIR EXPRESSION IN C.G.S. UNITS (ERGS) AND THE FACTORS BY WHICH THE FIGURES ARE OBTAINED

Therm for the supply of gas.

1 Therm = 100,000 British Thermal Units (B.Th.U.) or nominally gas enough to heat 1000 lb of water 100° F. $= 1\cdot05 \times 10^{15}$ ergs.

Board of Trade Unit of electrical energy.

1 B.T.U. (kilowatt-hour) $= 10^{10} \times 3600$ ergs $= 3\cdot6 \times 10^{13}$ ergs.

A hundred metres foot-race.

A man of 10 stone running 100 metres in 10 seconds.

10 stone = 140 lb. $= 6\cdot3 \times 10^4$ grammes.

$\frac{1}{2}mv^2 = \frac{1}{2} \times (6\cdot3 \times 10^4) \times 10^6$ g (cm/sec)$^2 = 3\cdot15 \times 10^{10}$ ergs.

More than a thousand such runners would be needed to express the energy of one B.T.U.

Horse-power-hour.

1 horse-power $= 7\cdot46 \times 10^9$ erg/sec. $= \frac{3}{4}$ kilowatt (approx.).

1 horse-power-hour $= 7\cdot46 \times 10^9 \times 3600 = 2\cdot69 \times 10^{13}$ ergs, three-quarters of a B.T.U.

One-ton motor car.

One-ton motor car ($1\cdot016 \times 10^6$ grammes) travelling at 30 mi/hr or 13·4 m/sec.

$\frac{1}{2}mv^2 = \frac{1}{2} \times (1\cdot016 \times 10^6) \times (13\cdot4 \times 100)^2 = 9\cdot1 \times 10^{11}$ ergs.

Forty would represent a B.T.U.

An ocean liner.

The projected *Queen Elizabeth* of 85,000 tons to travel at 28 knots.

1 knot = 0·51453 m/sec.

$\frac{1}{2}mv^2 = \frac{1}{2}\,(8\cdot5 \times 1\cdot016 \times 10^{10}) \times (1\cdot44 \times 10^3)^2$ ergs.

$= 8\cdot95 \times 10^{16}$ ergs = 2500 B.T.U.

Air-motion in a ten-metre layer over a square between consecutive isobars with intervals of 4mb.

Energy $= \frac{1}{8}\,\frac{hb^2}{\rho\omega^2}\,\text{cosec}^2\,\phi$.

Energy in lat. 50° $= \frac{1}{8}\,(10^3 \times (4 \times 10^3)^2 \times 1\cdot3^2)/(1\cdot2 \times 10^{-3} \times 7\cdot3^2 \times 10^{-10})$

$= 5\cdot3 \times 10^{20}$ ergs.

Energy in lat. 55° $= 4\cdot7 \times 10^{20}$ ergs.

At the equator, if consecutive straight isobars could be balanced by the pressure of wind velocity, the energy would be infinite.

In the example that we have displayed in fig. 97 the winds of which we have been considering the energy are represented by the simple case of east wind and west wind flowing in opposite directions with a ridge of high pressure between them; but the daily weather charts contain other examples of the distribution of isobars associated with winds which have their own quantity of energy of moving air in operation, and the energy in the several cases can be calculated in a manner similar to that which we have used in the case of the obstinate anticyclone. We have indeed chosen four examples in which the isobars suggest a triangular form, and in which south-west and north-west or south-east and north-east winds with their associated pressure-differences help to make up the condition represented on the map. The four triangular specimens, figs. 98–101, are represented by portions of the charts of weather in the northern hemisphere in the British Section of the Daily Weather Report. The energy represented by the association of the lines of the several charts is marked in the legend.

We have not had the opportunity of looking through all the copies of the Daily Weather Report that have reached us during the past sixty years so that our selection is necessarily somewhat haphazard and may perhaps be improved upon by those who are familiar with the history of weather as represented in the Daily Weather Report. The cases which we have selected represent a sort of triangle erected upon a west wind or an east wind, or dependent upon the east wind or west wind passing north of them.

The examples occur as the result of the insistence of the easterly or westerly wind on maintaining its flow in spite of the interference of the winds from other directions. We may use as typical the winds which we call East and West, in conjunction with the North-East and North-West and the South-East and South-West in which the intervening forces seem to express themselves; and so evident is this association in the maps that their representation qualifies for the name of triangular instead of circular or oval.

An interesting feature of our maps is a belt of easterly wind north of the dominating westerly. It helps, with the Arctic north-wester and the tropical south-wester, to form an energetic region of low pressure (fig. 99) and it is somehow derived from the eastern continental area and may be the result of the export of cold air which accumulates in the winter half year on the land of the northern hemisphere. Its importance

Triangular depression with base of west wind

Energy:
720 squares,
$10{,}000 \times 10^6$ B.T.U.

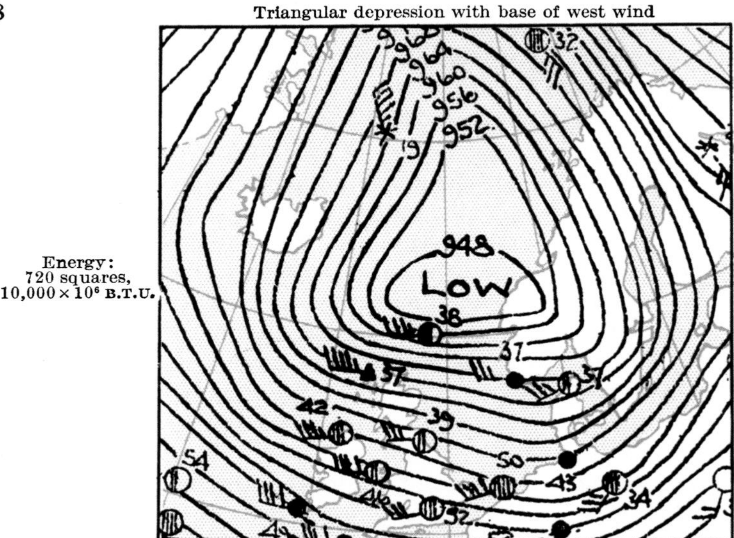

Fig. 98. From the chart for 29 January 1938, the westerly current agitated by a south-easter and a north-easter to form a triangular cyclone.

Triangular depression with cover of east wind

Energy:
266 squares,
3700×10^6 B.T.U.

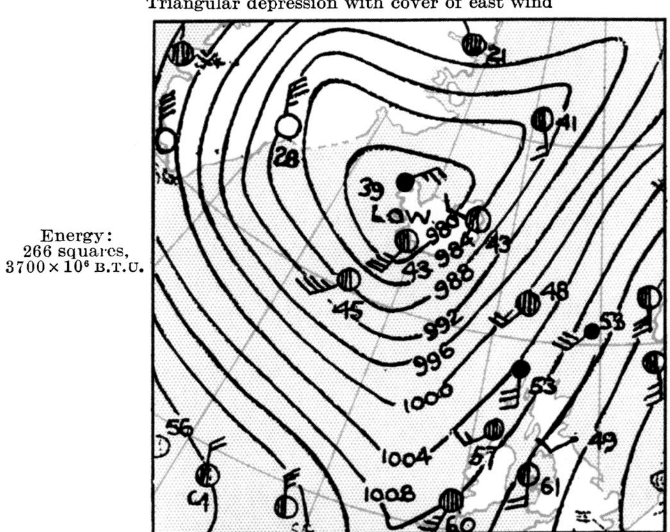

Fig. 99. From the chart for 30 September 1937. An easterly current in the north agitated by a north-wester on the west and a south-wester on the east to form a vigorous triangular cyclone centred over Iceland.

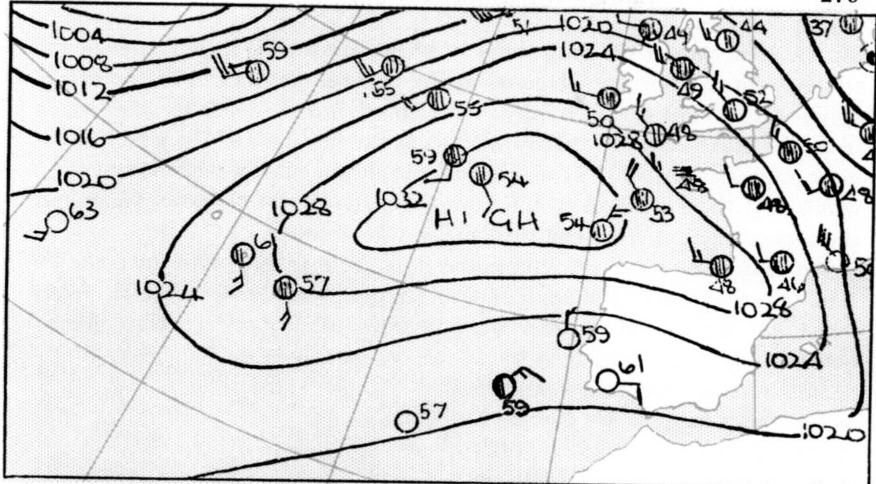

Fig. 100. From the chart for 28 March 1938. An anticyclone over the Bay of Biscay formed on the normal easterly wind over the Spanish peninsula with a south-wester over the Atlantic and a north-wester over Britain and France. Energy within 1024 mb: 42 squares, 600×10^6 B.T.U.

Anticyclone with cover of west wind

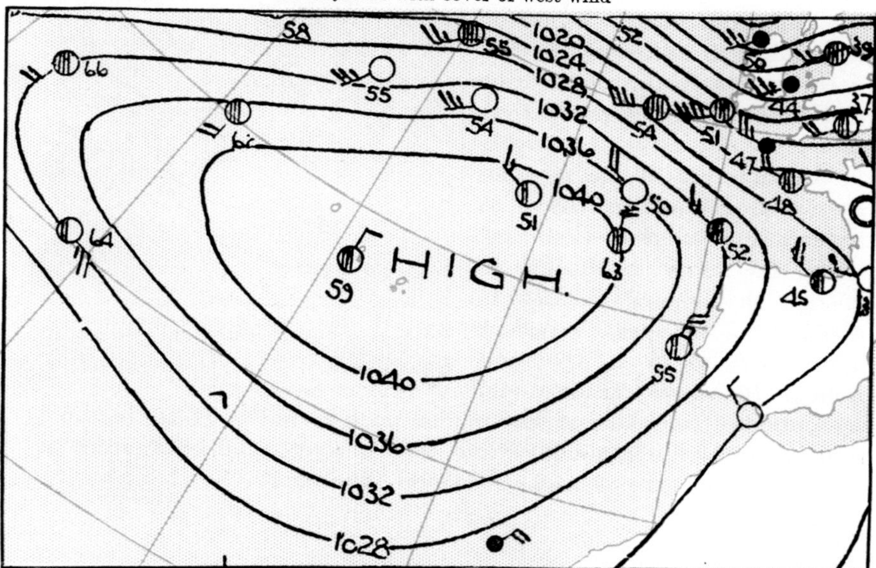

Fig. 101. From the chart for 28 January 1938, an anticyclone appended to the westerly current formed with the assistance of a south-easter on the western side and a north-easter on the eastern side. Energy within 1028 mb: 118 squares, 1600×10^6 B.T.U.

for the structure of the map leads us to remark on the importance of the winds from the northern regions in the general circulation of the atmosphere. It has not yet been fully investigated, but the persistence of winds from the north-east or north-west during the first half of the year 1938 is very impressive, and the question of the origin and persistence of such winds and their relation to the more fully investigated trade-winds and the return current from the west presses for solution.

We may take leave of depressions and their energy with an example that occurred while the text of this chapter was being written. After a brilliant March and an anticyclonic April and May our weather opened the first summer month with a depression that we might regard as typical of winter. Its picture is extracted, as a figure (fig. 102), from the chart for 6 p.m., 18 h, on 1 June of the International Section of the Daily Weather Report of the Meteorological Office. It was boisterous enough to destroy a good deal of the foliage that had been developed in the sunny spring and some of the full grown trees too. Its energy, estimated roughly in the same way as the previous examples, amounts to that of 68 squares in the steps of 4 mb each within the isobar of 1004 mb. In energy, that should amount to about 340×10^{20} ergs, equivalent approximately to a thousand million Board of Trade units with an electrical value of four million pounds sterling at 1*d.* per unit, four hundred million pounds if the moving layer could be regarded as covering a kilometre of height instead of 10 metres.

A boisterous 1st of June in England

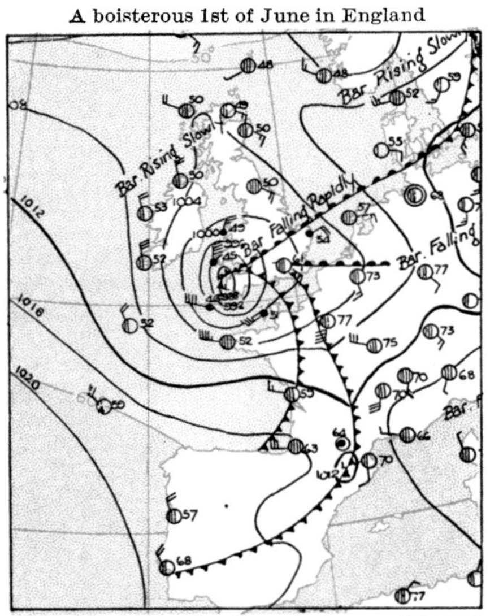

Fig. 102. The situation at 6 p.m. Energy within the closed isobar of 1004 mb is estimated at a thousand million Board of Trade units. Allowance should be made for curvature of the isobars.

CHAPTER VII

CHAPTER AND VERSE FOR WEATHER IN RELATION TO AGRICULTURE

Knowledge comes but wisdom lingers.
TENNYSON, *Locksley Hall*

They say that every why hath a wherefore.
DROMIO OF SYRACUSE in *The Comedy of Errors*

THE DAYLIGHT CYCLE

It is not unfair to say that vegetation on the earth's surface depends upon the use which is made of its daylight. The notable differences of agricultural practice are largely dependent upon the local daylight, not exactly as a physical quantity which can be accumulated or aggregated but as a sequence or cycle of opportunities that can be utilised. The importance of due regard for the influence of daylight has been expressed from time to time by Sir Daniel Hall, a former Director of the Rothamsted Experimental Station and now Director of the John Innes Horticultural Institution and chief Scientific Adviser of the Ministry of Agriculture. A lecture at the Royal Institution on *Light and Plants* by Professor V. H. Blackman, Director of the Biological Laboratories of the Imperial College, was published in *Nature* in 1936.

It is not always recognised that if one reckons daylight as the time between sunrise and sunset, taking no account of the small extension of the day by refraction, every place on the earth in the course of the year has the same aggregate amount of daylight, one half of the year. The poles take their two half years of darkness and light in turn; a place on the equator takes its share in half days, and places on intermediate latitudes take it in longer or shorter days according to the time of year.

Children from Kenya, where the sun always sets with singular promptitude at six o'clock, are apt to protest against being sent to bed in the daytime when they spend a summer in London.

If then daylight be indeed regarded as a controlling influence in agriculture, its fundamental coordinate, arrangements for the investigation

Daylight in relation to latitude

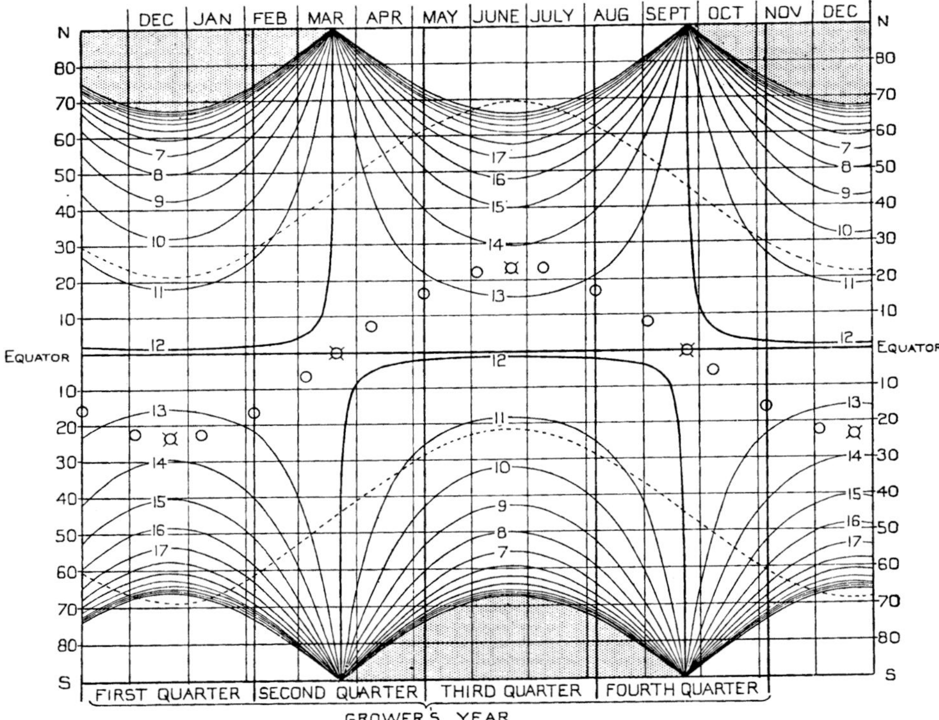

Fig. 103. Curves showing the latitudes of the places with a specified number of hours of daylight in the course of the year as reckoned by the succession of chapters of the Daylight Cycle for Agriculture, beginning with 6 November, or by the months of the civil year.

Horizontal lines are lines of latitude in steps of ten degrees north and south of the equator.

Thin vertical lines divide the year according to the months of the civil calendar. Thicker ones mark the quarters of the grower's year according to the Daylight Cycle, viz.: 6 November to 4 February, 5 February to 6 May, 7 May to 5 August, 6 August to 4 November with 5 November as audit day.

The row of circles crossing the equator shows the latitude at which the sun is vertical at noon, and parallel curves to the north and south mark the latitudes at which the altitude of the sun at noon is 45°. Then for a latitude *n* degrees *north* of the 45° line in the northern hemisphere the altitude of the sun at noon will be *n* degrees less than 45°. For the southern hemisphere a corresponding curve is shown.

The circles for the solstices and equinoxes are marked.

of the growth of plants should be so arranged as to keep in mind the position with regard to the duration of daylight. The reader may be interested in the author's attempt to put that principle into practice. It was explained in a paper, with *The Natural History of Weather* for its title, read before the Royal Meteorological Society on the longest day of 1934, after the reprint of this work in that year. The subject may deserve recognition as a chapter of this edition, perhaps as a chapter of chapters as will be apparent from its title.

We may introduce what we have to say by a diagram, fig. 103, borrowed from a recent edition of the second volume of the *Manual of Meteorology*, on which is set out a series of curves to enable the reader to trace for any latitude a curve of variation of its duration of daylight throughout the year, and to note the points at which the daylight reaches its maximum, the summer solstice, and its minimum, the winter solstice, June 21 or 22 and December 22 or 23 respectively, and also the two points of the year known as the equinoxes when in all latitudes the equality of length of day and night is in unison for one day with that at the equator. They are the two points where the length of daylight is changing most rapidly, the spring equinox, March 20 or 21, when its change is a notable increase from day to day in the Northern Hemisphere, and the autumn equinox, September 23 or 24, when the change from day to day is a notable decrease in daylight. The curves obtained in this way for latitude 30° N. and 60° N. are shown in another diagram, fig. 104, with the corresponding curve for London in latitude 51° 28′ N. In the lower part of the diagram the lapse of time is indicated by lines which separate the months of our civil calendar and in the upper part the same lapse is indicated by lines for successive weeks beginning from 6 November according to what we are calling the Daylight Cycle for Agriculture.

By an additional line on the diagram, fig. 103, it is possible for the reader to ascertain for himself the angular elevation of the sun at noon in any latitude and at any time of the year. The salient features of the daylight diagram are, of course, the solstices of summer and winter. The equinoxes take their turn as the points where the lines of equal duration of daylight converge in fig. 103, or the concurrent middle points of the nearly straight lines between the solstices in fig. 104. The equinox is marked in

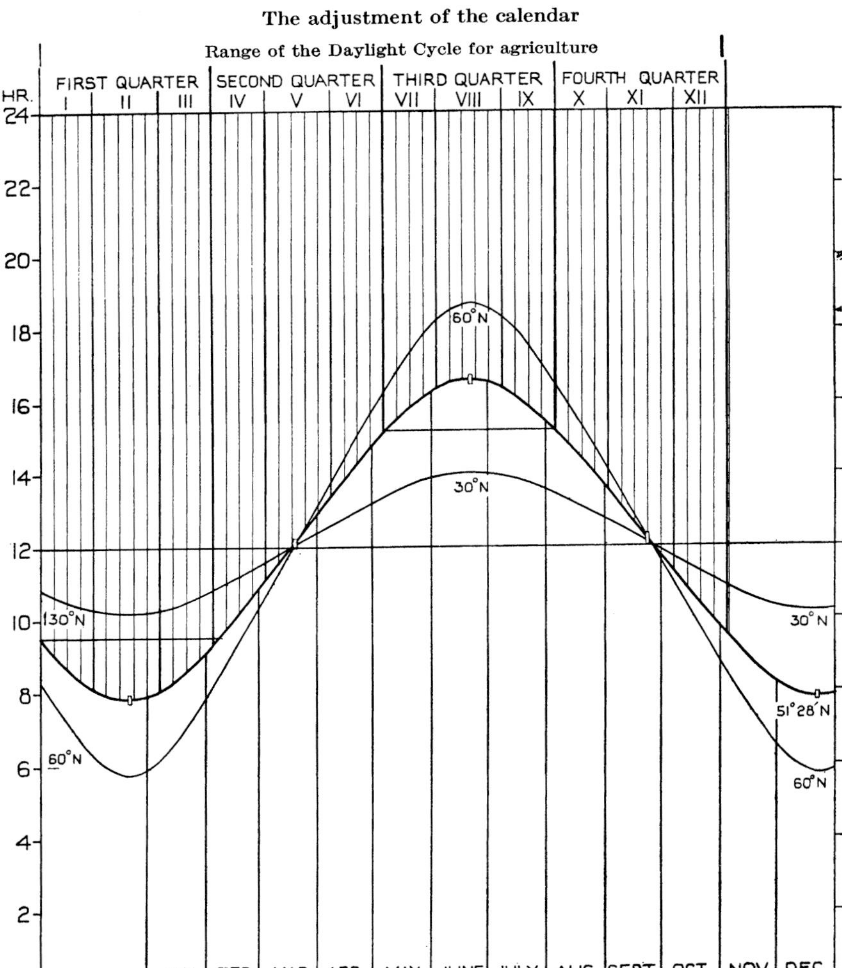

Fig. 104. Curves showing the number of hours of daylight in the latitude of London 51° 28′ N., also at 60° N. and 30° N. The invariable twelve hours a day at the equator is shown by the horizontal line through the mark of the hour.

the diagram by a rectangle of two days' width to allow for the unavoidable variation of date. We may arrange the divisions of our agricultural year with due regard to the four principal points. This implies the adoption of a special cycle for agricultural observations, and the paper referred to suggested a cycle based upon the week as the interval most appropriate for the purpose, intermediate between the day and the year.

The week was chosen by the Meteorological Council in consultation with the authorities of the Rothamsted Experimental Station for the *Weekly Weather Report* which was intended for the information of those interested in the growth of vegetation and has been continued for more than fifty years. The practice was to count everything in conventional weeks beginning with a Sunday, and so beginning each successive year one day earlier until it required fifty-three weeks to complete the year, and then begin again with fifty-two weeks.

The week generally makes a sufficient appeal in this country; but it is not so easily accepted on the Continent. There a period of five days is sometimes adopted, a half of ten days. That carries a suggestion of decimalisation which appeals to the scientific. For many years, when periodicity was the main object of research, the Meteorological Council published the results of their "first order" observatories as five-day means. In their Report for 1890 we read:

> "In future the publication will consist of a carefully arranged series of mean values, which it is believed will prove more generally useful than the publication of the hourly readings *in extenso*, which has hitherto taken place. The new series will comprise:
>
> (1) *Hourly* means of pressure, temperature, the difference between the dry and wet bulbs, and of wind-components, together with the amount and frequency of rain; these hourly means will be for periods of (*a*) five days, (*b*) calendar months, and (*c*) the year.
>
> From these will be computed the harmonic components of the hourly and yearly mean variations of pressure and temperature for each month and for the year."

There is, of course, a certain vulgarity about groups of days forming fractions of a year which are not in any sense decimal. But even with groups of five days we are not at the end of our difficulties, for in an ordinary year there are 73 five-day periods with a day over in leap year. Seventy-three numbers may be quite effective for the examination of a series of values in a table; but when it is desired to group them according to the duration of daylight the difficulty is too serious. An odd number of days is perhaps better than an even number because it has a middle day.

Another difficulty is concerned with the date of commencement of the cycle. The old *Weekly Weather Report* required each week to begin with its Sunday and so the first day of its year varied from the 28th of December to the 3rd of January. A more modern version of the *Weekly Weather Report* runs from 1 March 1936 to 27 February 1937, following one running from 3 March 1935 to 29 February 1936.

In the paper referred to it was thought better to adhere to seven days as the ordinary period without insisting upon the first being a Sunday, and group the "weeks" adequately (call them "verses" because they take their turn) with reference to the solstices and equinoxes, the cardinal points. The period allocated to each of these is a quarter of the year centred at a cardinal point; it may be expressed as thirteen verses for each. There would be an extra day to make up the three hundred and sixty-five for a year and two extra days for a leap year. For each quarter there would be thirteen verses of which four may form a first "chapter" with five for the chapter which carries a solstice or equinox, and four for the concluding chapter of the quarter.

The calendar month is, oddly enough, adopted without demur on an international basis for the purpose of grouping data of the weather for comparison between different countries. It seems very singular that it should be so, for the calendar month does not agree with the lunar period that carries the same name in astronomy and the calendar months are not even of equal length. In these conditions it is not surprising that remarks should be made about meteorology not being an exact science. It can hardly be so when our fundamental coordinate, time, has been so hopelessly confused for us, when it gives the same name to a period of 28 days and one of 31 days. There can be little doubt that astronomers would help us to put our time coordinate in better order. They have already given us the necessary particulars about daylight.

So in suggesting a daylight cycle for agricultural students of meteorology it was decided to adhere to the week, or perhaps more strictly speaking, to the "verse" of seven days, as the period intermediate between the day and the year, to divide the year into four quarters centred at the solstices and equinoxes, each quarter to consist of thirteen weeks to be regarded as "verses" in the sequence of weather, which could be grouped in chapters, of which the first and last would carry four verses each and the middle one of five verses should have a solstice or an equinox as nearly as possible on its middle day. "Chapters" need not all be of equal length.

That would account for 364 days and would leave one day over in each ordinary calendar year and two days over in leap year to be accounted separately. The final question is when should the cycle begin and what should be done with the additional day or two days? Retaining an extra day for each year on 5 November it might be worth while to reckon 6 May as an extra day every leap year and so keep in view the comparative characteristics of the beginning and middle point of our daylight cycle.

The cycle should begin when growth is in prospect rather than in being. Agriculture is busily occupied before the 1st of January. Dealing with the Northern Hemisphere it seemed desirable to consider which day, from an agricultural or meteorological point of view, was the least important in the year, and the choice fell upon the 5th of November for an ordinary year, which we celebrate with bonfires of rubbish, and the 4th and 5th of November in a leap year. So the New Cycle would begin on 6 November at a time when the year, after rapid loss of daylight since the autumn equinox, gives convincing signs of lapsing into winter. Then the seven-day verses might be counted from the date of commencement and not be called weeks as they do not necessarily begin with Sunday.

We may notice that in this cycle Christmas week is the eighth verse, the first day of our calendar year, 1 January, is the first day of our ninth verse, and the first day of July is the first day of our thirty-fifth verse, so that the relation of our verses to the weeks of the current year is not difficult to reckon.

Aggregates for comparison with Nature's agricultural integrals

We are illustrating the suggestion by five diagrams, figs. 105 to 109. We begin (fig. 105) with the aggregation week by week of Daylight, Sunshine, Rain and Warmth according to the experience of a Cambridge Farm during the Daylight Cycle 6 November 1932 to 5 November 1933 which we have dignified with the title of *Annus Mirabilis* because, according to a quotation from *The Times*, the corn crop was of outstanding merit.

For the next (fig. 106) we have in similar form the normals of the corresponding elements for the agricultural district England East as set out in the summary of the *Weekly Weather Report* for the period 1881–1915.

And in fig. 107 on p. 291 under the title of "Expectancy and Reality" we have the comparison of the weekly contribution from the Cambridge Farm 1932–3 with the normals of fig. 106 for the first half of the cycle and the available information about the direction and force of the winds.

Our addition in this edition includes two diagrams which give the aggregation of the normals for two districts of the British Isles, England East and Scotland East, with the corresponding data for the first half of the current Daylight Cycle, 6 November 1937 to 6 May 1938.

The selection of those two districts was originally made because on looking into the relation of meteorology and agriculture some years ago it was noticed that normally Scotland gave a better yield of barley than England and it was suggested that it might be on that account that the whisky industry was more closely associated with Scotland or sometimes with Ireland than with England. The data for the first half of the current cycle initiate a comparison of the *Annus Mirabilis* 1932 with the cycle 1937–8 which displays remarkable features. It may be noted in fig. 107 that in the corresponding period of 1932 at the Cambridge Farm the rainfall was obviously less than normal while the sunshine and warmth were greater than normal and the accumulated temperature below 42° was less than normal; whereas in fig. 108 we may notice that the rainfall for the first quarter was very nearly normal but in the second quarter there was so little that a failure of agricultural crops was generally anticipated; the sunshine was rather above normal and March was the warmest on record.

It will be interesting to continue the comparison, for the current cycle has disclosed some notable peculiarities since May 6. There was the introduction of summer by the storm of 1 June represented in fig. 102 which "attained a violence unprecedented for the time of year since

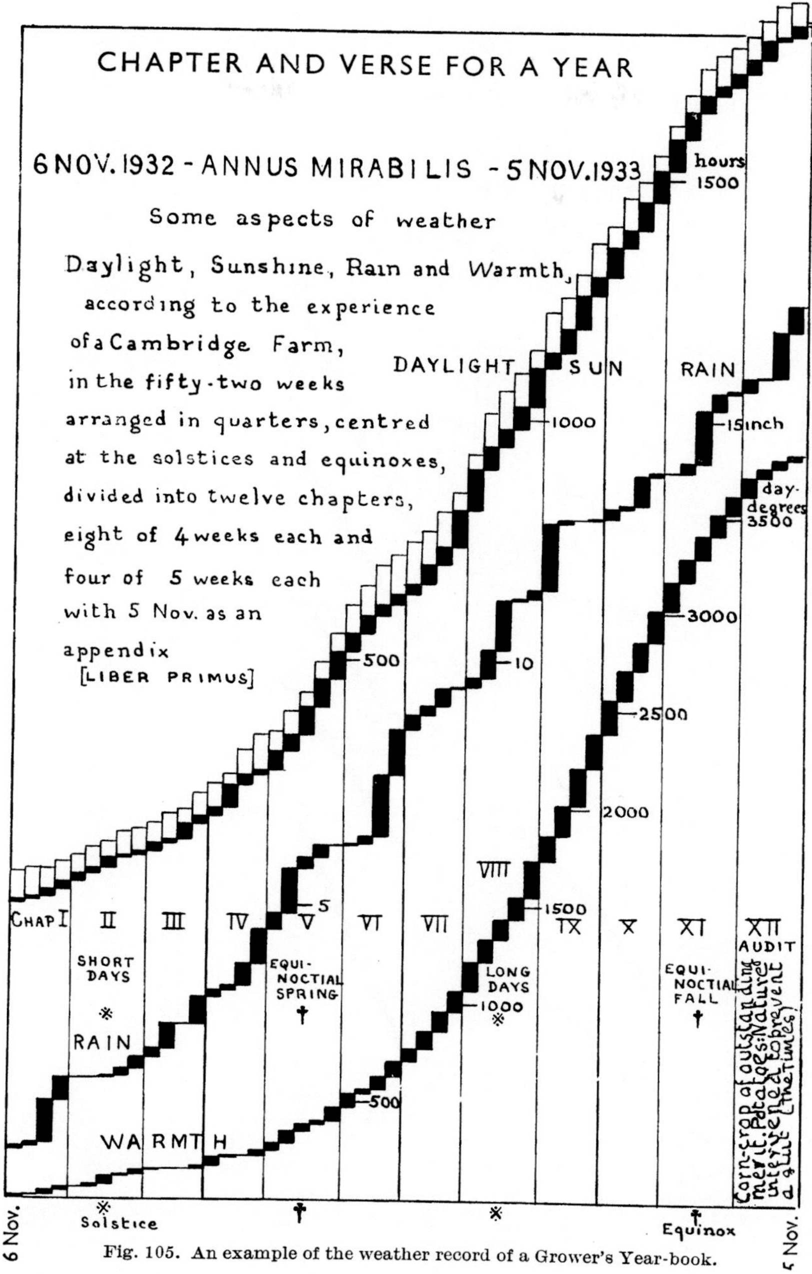

Fig. 105. An example of the weather record of a Grower's Year-book.

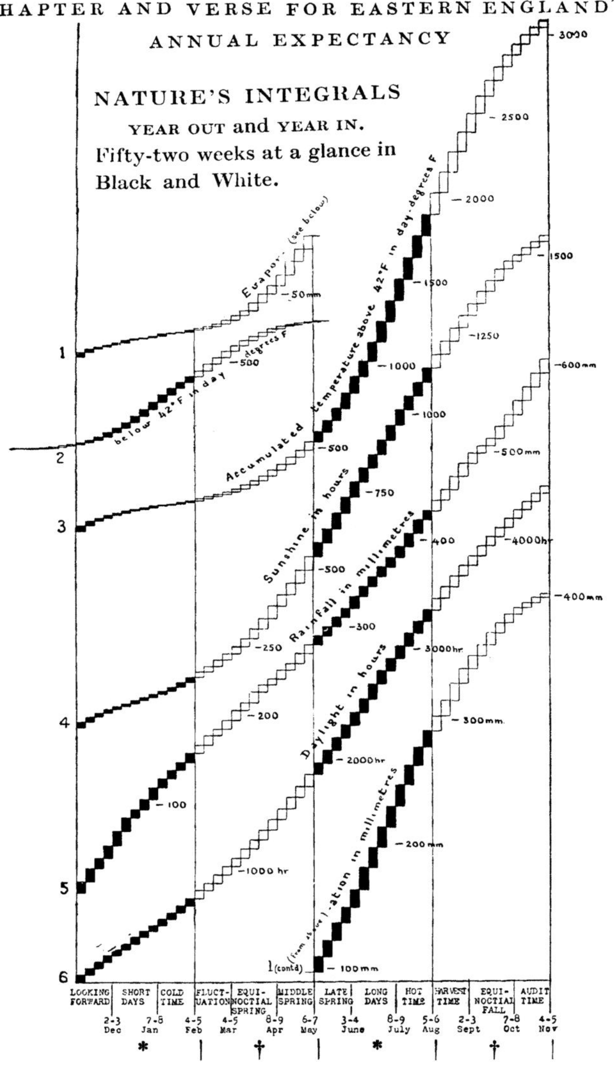

* Winter and Summer Solstice. † Spring and Autumn Equinox.

Fig. 106. Successive weekly amounts of Evaporation, Accumulated Temperature below 42° and above 42°, also of Daylight, Sunshine and Rainfall in the verses of the daylight cycle of which the chapters are marked at the foot of the diagram with the dates of their commencement in the civil calendar. The weekly contributions for the solstitial quarters are shown as black rectangles, those for the equinoctial quarters as white rectangles.

Weekly contributions in the half-year from 6 November 1932 ■
normals for England E □

The supplement below shows the sequence of wind-frequency in quadrants. The black part of a weekly column shows the frequency of winds of force 3 or over. The figure above the column is the Beaufort force of the highest wind of the week.

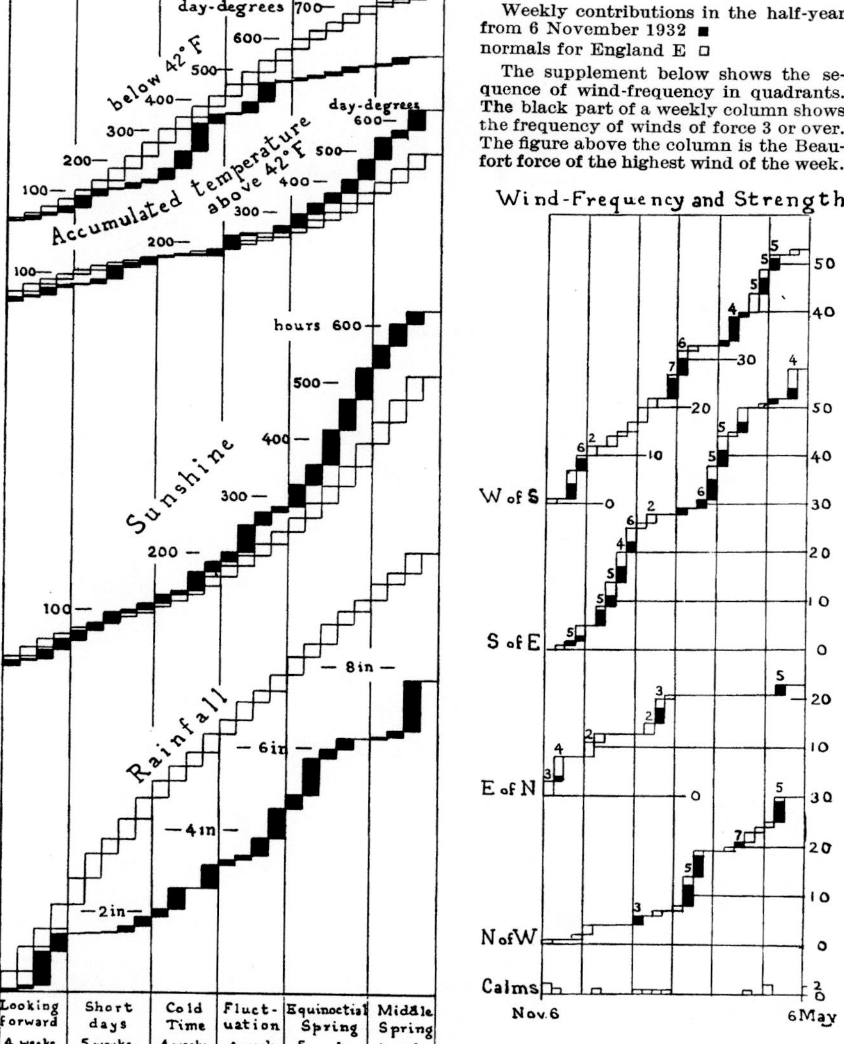

Fig. 107. Weekly contributions at Cambridge University Farm, 1932–3.

Cambridge 1937–8

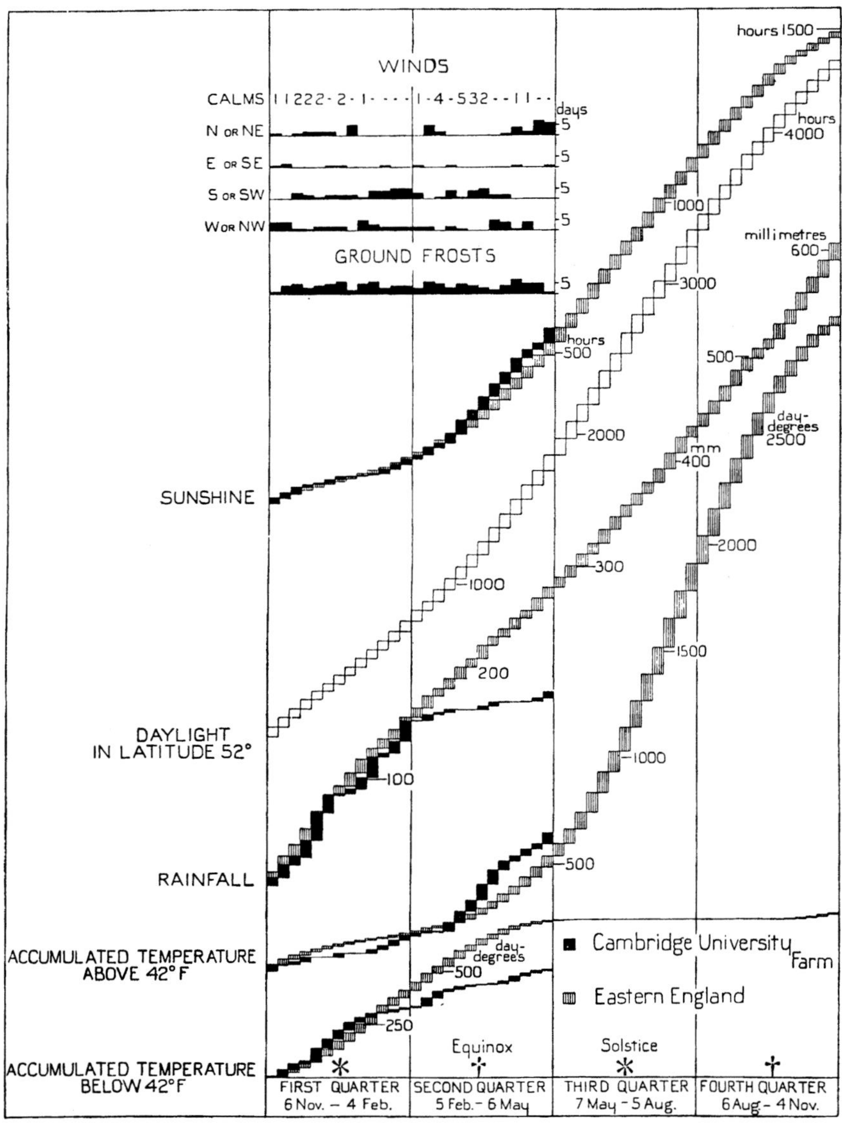

Fig. 108. The first half of the cycle of weather for agriculture 1937–8 at Cambridge University Farm with the District Normals for Eastern England.

Craibstone 1937–8

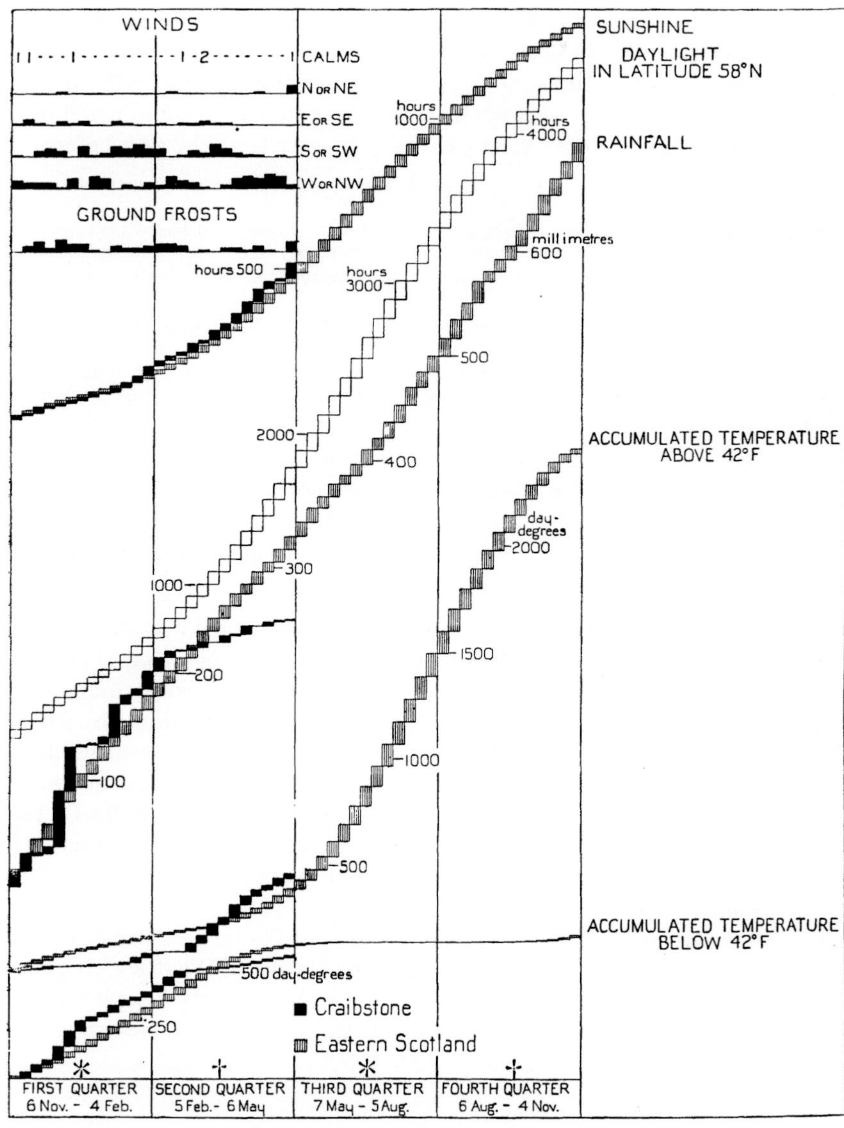

Fig. 109. The first half of the cycle of weather for agriculture 1937–8 at Craibstone with the District Normals for Eastern Scotland.

systematic wind-measurements began and possibly for a much longer period". "The last three days of June provided two depressions with gales almost as severe as those of 1st June." July was reasonable but has been followed by a number of days of thunderstorm, some with floods, particularly in the early part of August. We have culled some headlines from the newspapers; they refer generally to events of the day before.

The Times Aug. 2. 85° in London. Bank Holiday sunshine. West Country storms. Drenched crowds.

Aug. 5. Havoc of the storms. Over six inches of rain. Roads and Railways flooded.

Aug. 6. Storm damage on East Coast. Houses struck by lightning. Torrential rain.

Observer Aug. 7 (when it rained all day in Kensington). Storm's trail of damage. Lightning, floods and landslides.

The Times Aug. 12. Havoc of the storm. Lightning and flood. Snow in Yorkshire village.

The Times of Aug. 2 included a letter from a Borrowdale lady to say "our gauge has recorded 64 in. since January 1 of which 8·2 have fallen in the last 24 hours (July 30)".

So after some months of misgiving as to where our rain had gone to we are encouraged to revert to our inquiry as to where it comes from. The characteristic of the year for the casual observer has been the persistence of the motion of the clouds from the same direction for long periods.

* * * *

Interest in the numerical relations of weather and crops was stimulated by a paper read before the Royal Society in 1905 suggesting that the yield of wheat for Eastern England was more or less accurately indicated in advance by the amount of rainfall in the previous autumn, the drier the autumn the greater the crop. The comparison of data of weather and crops formed the subject of a paper before the Royal Statistical Society. In the early years of the present century it was developed by Mr R. H. Hooker in two presidential addresses of the Royal Meteorological Society, with a diagram illustrating the effect upon crops of the deviation of the meteorological elements from their normals.

So the examples of the working of our Daylight Cycle, figs. 108 and 109, may be regarded as a final contribution to the Drama of Weather with the hope that they may aid its readers to put the oldest of the sciences on a scientific footing and justify Ruskin's anticipation already quoted.

APPENDIX TO CHAPTER III

STEP-DIAGRAMS OF THE SEASONAL DISTRIBUTION OF RAINFALL IN ALL PARTS OF THE HABITABLE GLOBE AND OF THE CORRESPONDING CHANGES IN THE LOCAL ATMOSPHERIC LOAD AND OF THE TEMPERATURES OF THE DAY AND THE NIGHT

These diagrams are constructed from monthly records of rainfall at a large number of stations on the continents and islands of the world, the monthly averages of pressure and of maximum and minimum temperatures at sea as well as at land-stations.

The information about rainfall (fig. 62) is taken from the figures given in the publication of the Meteorological Office called the Réseau Mondial. For the pressure (fig. 63) the figures have been taken from observations at station-level at land-stations given in *World Weather Records* and for the sea from monthly charts of the distribution of pressure over the hemispheres.

For the temperatures (fig. 64) the step-diagrams carry a black column the top of which represents the normal highest temperature of the day (mean daily maximum) for each month of the year, and the bottom the normal lowest temperature of the night (mean daily minimum) for the month. The vertical boundaries of the columns have been carried down to the lowest monthly normal night temperature at the station, which with the maximum and the minimum can be read on a scale showing degrees Fahrenheit on one side and tercentesimal (Centigrade plus 273) on the other. A separate scale is provided for each zone of ten degrees of latitude on the map. The top line of the zones beyond 40° North or South indicates 300 tt (81° F), and 310 tt (99° F) for the other zones.

The information for the land-areas in each of the many step-diagrams is taken from the observations recorded for selected stations and endeavour has been made to give the reader a sufficient clue to the actual positions of the stations selected; but the diagram itself makes a considerable patch on the map and the identification of the point to which the patch refers is not easy. The attempt is on this wise—first of all the world is divided into zones of ten degrees of latitude. Then endeavour has been made to give figures for a station within every ten degrees of longitude of the zone, which would mean thirty-six stations for each zone. But there being no fixed stations on the sea only those on islands and continents within the zone are numbered.

The Réseau Mondial aspires to give the records from two stations from each ten-degree square. Our effort only runs to one, and even that aspiration is seldom fully satisfied; it is not always or even generally possible to choose the same stations for temperature and rainfall as well as pressure.

In the rainfall map the idea was to mark the position of the station by a black dot, put a number close to it and a corresponding number on the step-diagram belonging to it, putting the diagram somewhere in the margin of the map if there was no room for it in its own locality.

In the temperature map the number itself is used as sufficient to mark the position of the station and the position of the diagram is governed by the indication which has to be given of the scale appropriate to the zone. When no confusion is likely to be caused the reader is left to seek the proper record for himself; but when the selection seems dubious a corresponding number is fixed to the lower part of the diagram.

Readers may like to verify the positions of the selected stations in their own atlases, so here are given the names and numbers of the stations for which diagrams are mounted. They are arranged according to zones of ten degrees of latitude.

Index to the step-diagrams

Note. Pressure values for the sea have been read from monthly charts of mean isobars; the values are for the centre of each 10-degree square.

Names and longitudes of stations supplying information about rainfall ●, pressure p, and temperature tt, arranged in successive zones of 10 degrees of latitude.

* High level station above 1000 ft.

70–80° N

Station	Long.	●	p	tt	Station	Long.	●	p	tt
Barrow	156° W	—	—	2	Mehavn	28° E	6	—	—
Gaasefjord	89	—	—	3	Vardö	31	7	—	7
Upernivik	56	1	1	4	Malye Karmakouly	53	8	—	8
Jan Mayen	8	—	—	5	Waigatz	59	—	—	9
Spitsbergen	14° E	2	—	6	Dickson	80	—	—	10
Cap Thordsen	16	4	—	—	Sagastyr	124	9	—	1
Treurenberg Bay	17	3	—	—					
Gjesvaer	25	5	2	—					

60–70° N

Station	Long.	●	p	tt	Station	Long.	●	p	tt
Nome	165° W	1	1	3	Bodö	14° E	—	11	—
Tanana	152	—	2	4	Haparanda	24	—	12	15
Fairbanks	148	—	—	6	Helsingfors	25	9	—	16
Valdez	146	2	3	5	Kola	33	—	—	17
*Dawson	139	—	4	7	Archangel	41	—	13	18
*Carcross	135	3	—	—	Ust-Zylma	52	—	14	—
Fort Good Hope	129	—	—	8	Berezov	65	10	—	19
Hay River	115	—	—	9	Obdorsk	67	—	15	—
Godthaab	52	4	—	10	Sourgout	73	—	16	20
Jacobshavn	51	—	5	—	Doudinka	86	—	17	21
Ivigtut	48	—	6	—	Jakoutsk	130	11	18	1
Angmagsalik	38	5	7	11	Verkhoiansk	133	—	19	2
Stykkisholm	23	—	8	—	Markovo-on-Anadyr	171	—	20	—
Vestmanno	20	6	—	12	Novo-Mariinskii Post	178	12	—	—
Berufjord	14	—	9	—					
Thorshavn	7	7	10	13					
Christiansund	8° E	8	—	14					

50–60° N

Station	Long.	●	p	tt	Station	Long.	●	p	tt
St Paul Is.	170° W	—	1	—	Uccle	4° E	11	—	17
Dutch Harbour	166½	1	2	7	De Bilt	5	—	11	—
Kodiak	152	—	3	—	Copenhagen	13	—	12	—
Sitka	135	2	4	8	Potsdam	13	12	—	—
*Kamloops	120	3	5	9	Uppsala	18	13	—	—
*Edmonton	113½	—	6	10	Warsaw	21	—	—	18
*Prince Albert	106	—	7	—	Vilna	25	—	13	—
*Qu'Appelle	104	4	—	—	Leningrad	30	14	—	19
Port Nelson	93	5	—	11	Moscow	38	15	14	—
Moose Factory	80½	—	8	12	Saratov	46	—	—	20
Fort George	79	—	—	13	Kazan	49	16	15	—
*Mistassini Post	74	6	—	—	Orenburg	55	—	16	—
Belle Isle	55	7	—	14	Perm	56	—	—	21
Valentia	10	8	9	15	*Slatoust	59	—	—	22
Aberdeen	2	9	10	—	Bogoslowsk	60	—	—	23
Greenwich	0	10	—	16	Ekaterinburg	61	17	17	—

50–60° N (*continued*)

Station	Long.	●	p	tt	Station	Long.	●	p	tt
Tobolsk	68° E	—	—	24	*Nertchinskii Zavod	120° E	21	—	—
Omsk	73	18	18	—	Blagoveschensk	127½	22	23	4
Barnaul	84	—	—	25	Nikolaevsk-sur-Amour	141	23	24	—
Tomsk	85	19	19	—	Okhotsk	143	24	—	5
Jenisseisk	92	—	20	1	Petropavlovsk	159	—	25	6
*Irkutsk	104	20	21	2					
*Tchita	113½	—	22	3					

Sea: Temperature values are shown for the following 5-degree squares between 50 and 55° N: 180–175 W, 160–155, 150–145 W, 170–175 E.

40–50° N

Station	Long.	●	p	tt	Station	Long.	●	p	tt
Victoria	123° W	1	—	2	Marseille	5° E	—	11	—
Portland	123	—	1	—	*Zürich	9	12	—	—
*Spokane	117	—	2	—	Rome	12	13	—	12
*Salt Lake City	112	2	—	3	Vienna	16	14	—	—
*Cheyenne	105	—	3	4	Lesina	16	—	12	—
*Bismarck	101	—	4	—	*Sofia	23	15	—	13
Winnipeg	97	3	—	—	Bucharest	26	—	13	—
St Paul	93	—	5	—	Odessa	31	16	14	14
Port Arthur	89	4	—	—	Novorossiisk	38	17	—	—
Chicago	88	5	—	5	Astrakhan	48	18	15	15
Detroit	83	—	6	—	Krasnovodsk	53	19	16	16
Montreal	74	—	7	6	*Tachkent	69	20	17	17
Eastport	67	—	8	—	*Alma Ata	77	—	18	18
St John, N.B.	66	6	—	7	Mukden	123	21	19	—
Anticosti, SW pt.	64	7	—	—	Nikolsk Ussuriysky	132	—	20	—
St John's	53	8	9	8	Vladivostok	132	22	—	1
*Madrid	4	9	—	9	Ochiai	143	23	21	—
Nantes	2	10	10	10	Syana	148	24	—	—
Paris	2° E	11	—	11					

Sea: Temperature values for 5-degree squares 40–45° N: 140–145 W, 130–135, 50–55, 40–45, 30–35, 20–25 W, 150–155 E, 160–165 E.

30–40° N

Station	Long.	●	p	tt	Station	Long.	●	p	tt
San Francisco	122° W	1	1	4	*Algiers	3° E	12	12	14
San Diego	117	—	2	5	Catania	15	—	13	—
*Modena	114	2	—	—	Athens	24	—	14	—
*El Paso	106½	—	3	—	Alexandria	30	13	—	15
*Santa Fe	106	—	4	—	Cairo	31	14	—	—
*Denver	105	—	—	6	Nicosia	33	15	—	16
Little Rock	92	—	5	—	Beirut	35	—	15	—
St Louis	90	3	—	7	Baghdad	44	—	16	—
Mobile	88	4	6	8	Lenkoran	49	16	—	—
Washington	77	5	—	9	*Teheran	51	17	—	17
Hatteras	76	—	7	—	*Meshed	60	18	—	18
Bermuda	65	6	8	10	*Quetta	67	19	17	19
Flores	31	7	—	—	Lahore	74	—	18	—
Horta	29	8	—	11	*Simla	77	20	—	20
Ponta Delgada	26	—	9	—	Tientsin	117	21	19	1
Madeira	17	9	10	12	Shanghai	121	22	20	2
Lisbon	9	10	—	13	Nagasaki	130	23	—	—
Gibraltar	5	—	11	—	Tokio	140	24	21	3
Palma	3° E	11	—	—					

Sea: Temperature values for 5-degree squares 30–35° N: 180–175 W, 140–145, 130–135, 60–65, 50–55, 40–45, 30–35, 20–25, 10–15 W, 140–145 E, 150–155, 160–165, 170–175 E.

20–30° N

Station	Long.	●	p	tt
Midway Is.	177° W	1	—	5
Honolulu	158	2	1	6
Mazatlan	106	—	2	—
*Leon	102	3	—	—
*Tulancingo	98	4	—	—
Corpus Christi	97	—	3	—
Galveston	95	5	—	7
Havana	82	6	4	8
Nassau	77	7	5	9
*La Laguna	16	8	—	10
Insalah	2° E	[9]	—	11
Suva	25	—	—	12
Cairo (Helwan)	31	—	6	—
Aswan	33	[9]	—	13
Bushire	51	10	7	14
Jask	58	—	8	—
Robat	61	11	—	—

Station	Long.	●	p	tt
Hyderabad	68° E	—	9	—
*Jaipur	76	12	10	15
*Nagpur	79	13	—	—
*Darjeeling	88	14	—	—
Calcutta	88	15	11	16
Gauhati	92	16	—	—
*Cherrapunji	92	17	—	—
Akyab	93	—	12	—
Mandalay	96	18	—	1
Phu Lien	107	19	—	—
Moncay	108	—	13	—
Hong Kong	114	20	14	2
Taihoku	122	21	—	3
Santo Domingo	122	22	—	—
Naha	128	23	15	4
Titizima (Bonin Is.)	142	24	—	—

Sea: Temperature values for 5-degree squares 20–25° N: 160–165 W, 110–115, 60–65, 50–55, 40–45, 30–35, 20–25 W.

10–20° N

Station	Long.	●	p	tt
*Mexico, Tacubaya	99° W	1	1	—
*Oaxaca	97	—	—	4
*San Salvador	89	2	2	—
Belize	88	3	—	5
Jamaica	78	4	—	6
Port au Prince	72	5	3	—
*Caracas	67	6	—	7
San Juan	66	—	4	—
Barbados	60	7	—	8
St Vincent	25	8	—	9
Bathurst	17	9	—	10
Timbuktu	3	—	†	11
*Sokoto	5° E	10	—	—
*Kaduna Capital	7	11	—	—
*Hadeija	10	—	—	12

Station	Long.	●	p	tt
*El Fasher	25° E	—	—	13
*Khartoum	33	12	5	14
Berbera	45	13	—	15
Aden	45	—	6	—
Bombay	73	14	7	16
Cochin	76	15	—	17
*Bangalore	78	16	8	—
Madras	80	17	—	18
Waltair	83	18	9	—
Port Blair	93	19	—	1
Rangoon	96	20	10	2
Nhatrang	109	21	11	—
Bolinao	120	22	—	—
Manila	121	23	12	3
Guam	145	24	—	—

Sea: Temperature values for 5-degree squares 10–15° N: 135–140 E, 145–150 E.

† A diagram has been added to the pressure chart with no number.

0–10° N

Station	Long.	●	p	tt
Fanning Is.	159° W	1	—	8
Colon	79	2	1	—
*Bogota	74	3	—	9
El Peru	62	4	—	10
Georgetown	58	—	2	—
Paramaribo	55	5	—	—
Cayenne	52	6	—	11
Conakry	14	7	—	—
Freetown	13	—	3	—
Accra	0	8	—	12
Lagos	3° E	—	4	—
Zungeru	6	9	—	13
Libreville	9	10	—	14
Yola	12½	11	—	—
*Wau	28	12	—	—

Station	Long.	●	p	tt
*Mongalla	32° E	13	—	15
*Entebbe	32	14	5	16
*Harrar	42½	—	—	17
Minicoy	73	15	—	—
Colombo	80	16	6	18
Trincomali	81	17	7	19
Kota Radja	95	18	—	—
Penang	100	19	—	1
Singapore	104	20	—	2
Sandakan	118	21	—	3
Iwahig	119	22	—	4
Menado	125	23	—	5
Yap	138	24	—	6
Uyelang	161	—	—	7

Sea: Temperature values for 5-degree squares 0–5° N: 50–55 E, 60–65 E.

0–10° S

Station	Long.	●	p	tt	Station	Long.	●	p	tt
Malden Is.	155° W	1	1	8	Dâr-es-Salaam	39° E	—	—	17
Guayaquil	80	—	—	9	Lamu	41	11	—	18
Manaos	60	2	—	10	Seychelles	55	12	—	19
Taperinha	54	—	—	11	Padang	100	13	—	—
Para	48	—	—	12	Batavia	107	14	4	1
Turyassu	45	3	—	—	Pontianak	109	15	—	—
Barra do Corda	45	4	—	13	Pasoerocan	113	16	5	2
Quixeramobim	39	5	—	14	*Kajoemas	114	17	—	—
Pão dé Assucar	37	6	—	—	Ambon	128	18	—	3
Recife	35	—	2	—	Manokwari	134	19	—	4
Fernando Noronha	32	7	—	—	Daru	143	20	—	5
Ascension	14	—	—	15	Port Moresby	147	21	6	6
Loanda	13° E	8	—	16	Rendova	157	22	—	—
*Brazzaville	15	9	—	—	Tulagi	160	23	—	—
*Nairobi	37	10	—	—	Ocean Is.	170	24	—	7
Zanzibar	39	—	3	—					

Sea: Temperature values for 5-degree squares 5–10° S: 60–65 E, 70–75, 80–85, 90–95 E.

10–20° S

Station	Long.	●	p	tt	Station	Long.	●	p	tt
Samoa	172° W	1	1	7	*Salisbury	31° E	—	4	—
Niue Is.	170	2	—	8	*Zomba	35	14	—	17
Tahiti	150	—	—	9	*Antananarivo	48	15	5	—
Makatea	148	3	—	—	Tamatave	49	—	—	18
*Arequipa	72	4	—	10	Cocos Is.	97	16	—	1
Puerto de Arica	70	5	—	11	Christmas Is.	106	17	—	2
*Sucre	65	6	—	—	Broome	122	—	—	3
Cuyaba	56	7	2	12	Derby	124	18	—	—
*Bello Horizonte	44	8	—	—	Darwin	131	19	6	4
*Caetite	43	9	—	13	Daly Waters	133	20	—	—
Ondina (Bahia)	39	10	—	14	Mein	143	21	—	—
Aracaju	37	11	—	—	Harvey Creek	146	22	—	—
*St Helena	6	12	3	15	Samarai	151	23	—	5
*Gwelo	30° E	13	—	16	Suva, Fiji	178	24	—	6

Sea: Temperature values for 5-degree squares 15–20° S: 50–55 E, 60–65, 70–75, 80–85; 10–15 S: 110–115 E.

20–30° S

Station	Long.	●	p	tt	Station	Long.	●	p	tt
Rarotonga	160° W	1	—	8	Mauritius	58° E	13	—	18
Mangareva	135	—	—	9	Carnarvon	114	—	—	1
Easter Is.	109	2	—	10	Onslow	115	14	—	—
Punta Tortuga	71	3	—	—	Peak Hill	119	15	—	—
Iquique	70	4	—	—	*Nullagine	120	16	—	—
*Catamarca	66	—	1	12	*Laverton	122	17	—	2
*Salta	65	5	—	11	*Alice Springs	134	—	7	3
Goya	59	—	2	—	William Creek	136	18	—	—
Asuncion	58	6	—	13	Thargomindah	144	19	—	—
Rio de Janeiro	43	7	3	14	*Mitchell	148	20	—	—
*Windhoek	17° E	8	—	15	Rockhampton	150½	21	—	4
*O'Okiep	18	—	4	—	Brisbane	153	22	8	5
*Kimberley	25	9	5	16	Ouitchambo	166	23	—	—
*Bulawayo	29	10	—	—	Lifou	167	—	—	7
Durban	31	11	6	—	Norfolk Is.	168	24	—	6
Lourenço Marques	33	12	—	17					

Sea: Temperature values for 5-degree squares 25–30° S: 60–65 E, 70–75, 80–85, 90–95, 100–105.

Station	Long.	●	p	tt	Station	Long.	●	p	tt
				30–40° S					
*Juan Fernandez	79° W	1	—	8	*Katanning	118° E	12	—	—
Valdivia	73	2	—	9	*Coolgardie	121	13	—	—
Valparaiso	72	3	—	10	*Kalgoorlie	121½	—	—	2
*Santiago	71	4	1	—	Eucla	129	14	—	—
*Cordoba	64	5	—	—	Streaky Bay	134	15	—	—
Bahia Blanca	62	6	2	—	Adelaide	139	16	6	3
Buenos Ayres	58	7	3	11	Melbourne	145	17	—	4
Cape Town	18° E	8	4	12	Bourke	146	18	—	—
Heidelberg	21	9	—	—	Sydney	151	19	7	5
Port Elizabeth	26	—	5	—	Lord Howe Is.	159	20	—	—
East London	28	10	—	13	Auckland	175	21	—	6
Perth	116	11	—	1	Napier	177	22	—	7
				40–50° S					
Chatham Is.	177° W	1	—	4	Launceston	147° E	6	—	—
Isla Huafo	75	2	—	—	Hobart	147	7	—	1
Punta Galera	74	3	1	5	Invercargill	168	8	—	2
Sarmiento	69	4	—	—	Dunedin	171	9	—	—
Trelew	65	—	—	6	Christchurch	173	10	—	—
Puerto Madryn	65	5	2	—	Wellington	175	11	3	3
Kerguelen	70° E	—	—	7					
				50–60° S					
Islota de los					Año Nuevo	64° W	6	—	4
Evangelistas	75° W	1	1	2	Stanley	58	7	—	5
Punta Arenas	71	2	—	—	Cape Pembroke	58	—	3	—
Punta Dungeness	68	3	—	—	S. Georgia	37	8	4	6
Santa Cruz	68	4	2	—	Macquarie Is.	159° E	—	—	1
Ushuaia	68	5	—	3					
				60–70° S					
Port Charcot	63° W	1	—	—	S. Orkneys	45° W		1	1

List of Mountains in fig. 40

The mountains which are portrayed at the foot of fig. 40 (p. 114), reading from left to right are as follows:

Mt Pico (Azores)
Ben Nevis
Mt Kosciusko (Australia)
Greenland
The Antarctic Continent
Mt Etna
Fujiyama
Mt Erebus
Pike's Peak (U.S.A.)
Mt Blanc
Ararat
Popocatepetl (Andes)
Charles Louis (New Guinea)
Elbruz (Caucasus)
Mt Logan (Yukon)
Kilima Njaro (East Africa)
Mt McKinley (Alaska)
Everest
Khan Tengri (Turkestan)
Aconcagua (Argentine)

Note to p. 184

Attention has been called to some uncertainty about the lowest reading quoted for Verkhoiansk. The lowest reading is one of unavoidable uncertainty. Many years ago the late J. H. Poynting raised the question of the limit that might be reached by a thermometer contained in a well-insulated cup exposed to the open sky on a calm night, and the provisional answer seemed to be that there was no limit, or perhaps that it was a natural curiosity without much meteorological significance. And so it remains.

INDEX